ERICH HÜCKEL (1896–1980)

BOSTON STUDIES IN THE PHILOSOPHY OF SCIENCE

VOLUME 283

For further volumes:
http://www.springer.com/series/5710

ERICH HÜCKEL
(1896 – 1980)

From Physics to Quantum Chemistry

ANDREAS KARACHALIOS

Translated by
Ann M. Hentschel

Springer

Andreas Karachalios
Ernst-Reuter-Str. 14
55130 Mainz
Germany
karachal@uni-mainz.de

ISBN 978-90-481-3559-2 e-ISBN 978-90-481-3560-8
DOI 10.1007/978-90-481-3560-8
Springer Dordrecht Heidelberg London New York

Library of Congress Control Number: 2009939498

Printed on acid-free paper

Springer is part of Springer Science+Business Media (www.springer.com)

To the memory of my friend, Thanos Liarommatis

Acknowledgments

It is a pleasure for me to thank those persons who lent me their support.

David Rowe and Moritz Epple offered guidance and commentary during the growing stages of this work. I thank them heartily for their inspiring ideas and active assistance.

I am also especially grateful to W.H. Eugen Schwarz, Hans Sillescu, and Joachim Stocklöv for their valuable and helpful comments.

Finally, I'd like to thank the following persons for their efforts: Regina Grimm-Karachalios, Ann M. Hentschel, Peter Hülsmann and Dr. Volker Remmert.

Contents

Introduction

Emerging disciplines in the border zone between physics and chemistry have attracted the attention of historians of science particularly in the last 20 years.[1] Quantum chemistry,[2] as an offshoot of theoretical chemistry, has recently acquired some importance in the history of chemistry.[3] It is the product of close

[1]Cf. Hiebert, E.: *Discipline Identification in Chemistry and Physics*, in: Science in Context, 9(2) (1996), 93–119; Nye, M. J.: *Physics and Chemistry: Commensurate or Incommensurate Sciences?* in: *The Invention of Physical Science, Intersections of Mathematics, Theology and Natural Philosophy since the Seventeenth Century – Essays in Honor of Erwin N. Hiebert.* Kluwer Academic Publishers, Dordrecht 1992; *From Chemical Philosophy to Theoretical Chemistry: Dynamics of Matter and Dynamics of Disciplines, 1800–1950.* University of California Press, Berkeley 1994; Servos, J. W.: *Physical Chemistry from Ostwald to Pauling, the Making of a Science in America.* Princeton University Press, New Jersey 1990; *Chemical Sciences in the 20th Century: Bridging Boundaries*, edited by Carsten Reinhard. Wiley-VCH, Weinheim 2001 (incl. a comprehensive bibliography).

[2]In an earlier article I point out that the term "quantum chemistry" [*Quantenchemie*] first appeared in 1929. To my knowledge it was coined by the physicist Arthur Haas. Talks he had delivered before the Viennese Chemico-Physical Society in the spring of 1929 are assembled in his book: *Die Grundlagen der Quantenchemie: Eine Einleitung in vier Vorträge.* It was published by the Akademische Verlagsgesellschaft in Leipzig. See Karachalios, A: *Die Entstehung und Entwicklung der Quantenchemie in Deutschland*, in: Mitteilungen der Gesellschaft Deutscher Chemiker Fachgruppe Geschichte der Chemie, 13 (1997), 163–179, footnote 2.

[3]Cf. Gavroglu, K. and Simões, A.: *The Americans, the Germans and the Beginnings of Quantum Chemistry: The Confluence of Diverging Traditions*, in: Historical Studies in the Physical and Biological Sciences, 25 (1994), 47–110; *One Face or Many? The Role of Textbooks in Building the New Discipline of Quantum Chemistry*, in: *Communicating Chemistry. Textbooks and Their Audiences, 1789–1939*, edited by Lundgren, A. and Bensaude-Vincent, Bernadette. Science History Publications, USA 2000, 415–449; Simões, A.: *Converging Trajectories, Diverging Traditions: Chemical Bond, Valence, Quantum Mechanics and Chemistry, 1927–1937*, Ph.D. Thesis, University of Maryland, College Park, 1993; Simões, A. and Gavroglu, K.: *Different Legacies and Common Aims: Robert Mulliken, Linus Pauling and the Origins of Quantum Chemistry*, in: *Conceptual Perspectives in Quantum Chemistry*, edited by Calais, J. L. and Kryachko, E. S. Kluwer Academic Publishers, Dordrecht 1997; *Quantum Chemistry in Great Britain: Developing a Mathematical Framework for Quantum Chemistry*, in: Studies in the History and Philosophy of Modern Physics, 31B (2000), 511–548; *Issues in the History of Theoretical and Quantum Chemistry, 1927–1960*, in: *Chemical Sciences in the 20th Century: Bridging Boundaries*, edited by Carsten Reinhardt. Wiley-VCH, Weinheim 2001, 51–74; Simões,

interaction between chemistry, quantum physics and mathematics. Quantum chemistry applies the approximative procedures used in quantum mechanics to chemical problems, with chemistry being its main foundation. Its interdisciplinary character derives from physical bases that are remolded to suit issues in chemistry.[4] The methods, theories and concepts used in quantum chemistry characteristically transgress the boundaries between chemistry and quantum physics. This cross-cultural relationship is also reflected in the home fields of the scientists involved. It is primarily due to quantum physics that a reorientation of scientific inquiry to quantum-chemical issues emerged. Quantum physicists innovatively identified concepts and methods rooted in their own field that were of relevance to chemistry.

Germany was one of the most important centers for the developing field of quantum chemistry. It owed this distinction primarily to advances made in theoretical physics and quantum mechanics in the closing half of the 1920s. During that and the following decade communications between scientists in Germany, the United States and Britain engendered new contributions to the theory, along with attendant controversies between the individual schools of thought. A comparative analysis of the interrelated British and American developments illuminates the characteristic developments of quantum chemistry in Germany.

In Germany, too, most of the pioneers of the forming discipline were physicists attempting to understand chemical phenomena – initially more qualitatively than quantitatively – by means of physical concepts.[5] Erich Hückel is prominent among the contributors to the quantum theoretical foundations of organic chemistry. His is a familiar name among chemists and students today, reappearing frequently in current articles and monographs on aromaticity as well as in textbooks on organic chemistry, physical chemistry and quantum chemistry. "Hückel's $4n + 2$ rule" as

A.: *Chemical Physics and Quantum Chemistry in the Twentieth-Century*, in: *Cambridge History of Science* (8 Vols.), Volume 5, *Modern Physical and Mathematical Sciences*, edited by Mary Jo Nye. Cambridge University Press, Cambridge 2003, 394–412; Karachalios, A.: *On the Making of Quantum Chemistry in Germany*, in: Studies in the History and Philosophy of Modern Physics, 31B (2000), 493–510; *Giovanni Battista Bonino and the Making of Quantum Chemistry in Italy in the 1930s*, in: *Chemical Sciences in the 20th Century: Bridging Boundaries*, edited by Carsten Reinhardt. Wiley-VCH, Weinheim 2001, 75–104; Nye, M. J.: *From Chemical Philosophy to Theoretical Chemistry: Dynamics of Matter and Dynamics of Disciplines, 1800–1950*. University of California Press, Berkeley 1993, Cap. 9 Quantum Chemistry and Chemical Physics, 1920–1950, 227–261; Park, B. S.: Computations and Interpretations: The Growth of Quantum Chemistry, 1927–1967, Dissertation, Baltimore, Maryland, May, 1999 (UMI Microfilm 9927141).

[4]Cf. Gey, E.: *Der Zusammenhang von interdisziplinären Forschungssituationen und Kooperationsverhalten bei der Bildung und Entwicklung neuer Spezialgebiete, dargestellt am Beispiel der Quantenchemie*, in: *Interdisziplinarität in der Forschung: Analysen und Fallstudien*, hrsg. von Heinrich Parthey und Klaus Scheiber. Akademie-Verlag, Berlin 1983, 151–175; *Quantenchemie – eine interdisziplinäre Entwicklung*, in: Wissenschaft und Fortschritt 34 (1984) 11, 282–285; Kragh, H.: Quantum Interdisciplinarity: Friedrich Hund and Early Quantum Chemistry, unpublished paper delivered at the Georg-August-Universität in Göttingen, 6 February 1996, in honour of the 100th birthday of Friedrich Hund. I thank Helge Kragh for the transmission of the manuscript.

[5]Cf. Hund, F.: *Frühgeschichte der quantenmechanischen Behandlung der chemischen Bindung*, in: Angewandte Chemie 89 (1977), 89–94, p. 89.

well as the "Hückel molecular orbital theory" (HMO theory) for π-electron systems have immortalized his name, which also appears in conjunction with that of his mentor, Peter Debye, in the context of the "Debye-Hückel theory" of electrolytes.

Current histories of chemistry or quantum chemistry[6] take scant notice of Hückel's scientific profile. Shigeaki Kikuchi, for example, goes into great critical detail about the transition from Eugen Bamberger's classical "hexacentric valence system" to "Hückel's rule" and also addresses the related controversies. Yet he scarcely alludes to Hückel 's research as a quantum chemist.[7] William H. Brock, author of one the latest histories of chemistry, concedes a few sentences to Hückel's contribution to quantum chemistry in a brief chapter discussing the concept of "aromaticity."[8] Buhm Soon Park, on the other hand, sketches Hückel's contributions to the quantum theory of double bonding and the benzene problem in his very instructive dissertation on *Computations and Interpretations: The Growth of Quantum Chemistry, 1927–1967*.[9] Yet he emphasizes "Hückel's computational scheme" and interpretation rather than his methodology and conceptual approach. In 1996 Jerome A. Berson published a paper in *Angewandte Chemie* with the fitting title: "Erich Hückel – a pioneer of organic quantum chemistry: Life, work and tardy recognition."[10] In this German paper Berson reports on Hückel's scientific career and his contributions to quantum chemistry less from the point of view of a historian of science than from that of an organic chemist. He describes in particular how Hückel's concepts generally figured in research on chemical theory before and after World War II. The detailed necrology by Hermann Hartmann and Christopher Longuet-Higgins is also worth mentioning.[11] Although pertinent to Hückel's contribution to quantum chemistry, it offers a rather superficial panoramic view. Furthermore, there appeared two years ago a valuable paper by the historian

---

[6]Cf. footnote 3

[7]Kikuchi, S.: *Aromatic Seven-Membered Ring Compounds and Quantum Chemistry – The First Verification of Hückel's Rule and Its Influence*, in: Kagakusi Kenkyu. Journal of History of Science, Japan Series II, 29(174) (1990), 65–73; *Development of the Electronic Formula of Benzene*, in: Kagakusi Kenkyu. Journal of History of Science, Japan Series II 29(176) (1990), 209–218; *A History of the Structural Theory of Benzene – The Aromatic Sextet Rule and Hückel's Rule*, in: Journal of Chemical Education 74 (1997), 194–201.

[8]Brock, W. H.: *The Norton History of Chemistry*. W. W. Norton & Company, New York 1992, Chapter 14 Structure and Mechanism in Organic Chemistry; Aromaticity.

[9]Park, B. S.: *Computations and Interpretations: The Growth of Quantum Chemistry, 1927–1967*, Dissertation submitted to the Johns Hopkins University, Baltimore, Maryland May, 1999. (UMI: 9927141), 157–178.

[10]Berson, J. A.: *Erich Hückel – Pionier der Organischen Quantenchemie: Leben, Wirken und späte Anerkennung*, in: Angewandte Chemie 108 (1996), 2923–2937; *Erich Hückel, Pioneer of Organic Quantum Chemistry: Reflections on Theory and Experiment*, in: Angewandte Chemie International Edition in English 35 (1996), 2750–2764. *Chemical Creativity. Ideas from the Work of Woodward, Hückel, Meerwein, and Others*. Wiley-VCH, Weinheim 1999, Chapter 3 Erich Hückel and the Theory of Aromaticity: Reflections on Theory and Experiment, 33–75.

[11]Hartmann, H. and Longuet-Higgins, H. C.: *Erich Hückel, 9 August – 16 February 1980*, in: Biographical Memoirs of Fellows of the Royal Society 28 (1982), 153–162.

of science Helge Kragh,[12] in which Hückel's intellectual development as assistant to Max Born and Peter Debye in the 1920s is elaborated in great depth.[13]

The present book takes up where the aforementioned studies have left off. Its aim is to draw a more complete picture of Erich Hückel's important research efforts, in order to better assess his path-breaking papers on quantum chemistry. The focus is on Hückel's innovative amalgamation of organic chemistry with quantum physics. At the same time I augment this portrait with a glance at the social relations within the contemporary scientific community. By this I mean Hückel's exchanges with quantum physicists as well as organic and physical chemists during the 1930s.

A broad range of sources has been consulted for this study. It is based on Hückel's original research papers, an extensive interview and his autobiography. Hückel does not, incidentally, reveal very much at all about the initial conception and development of his own scientific ideas. The "Memoirs" by his brother Walter Hückel have also been taken as a source. I emphasize here, though, that such autobiographical documents cannot necessarily be taken as objective sources of information. The subjectivity of an autobiography is only embellished by the identity of author and protagonist as well as by an ambivalence between narration and real experience. Hückel himself concedes in his preface that his autobiography is just a retrospective on his past life, as it appeared to him at the age of 79. Nevertheless, Erich Hückel's autobiography and his brother's memoirs are good points of departure in my quest for documentation on the odd detail, neglected fact or circumstance.

The available material from Hückel's papers were examined with this goal in mind. His personal files at Marburg and the Rockefeller Archive Center in New York were also consulted along with other pertinent archival material. The correspondence with Max Born, Peter Debye, Werner Heisenberg, Friedrich Hund, Linus Pauling and the organic chemist Friedrich Richter have thereby taken center stage. The last-mentioned exchange was particularly informative for this historical study. At a time when many leading chemists with an interest in organic substances were rather skeptical of the new field of quantum chemistry, this communicative exchange of letters between Richter and one of the founders of the new subdiscipline

[12] Kragh, H: *Before Quantum Chemistry: Erich Hückel and the Physics-Chemistry Interface*, in: CENTAURUS 43 (2001), 1–16.

[13] A few other sketches of Hückel's life and work raise the false impression that his research has been thoroughly appreciated. However, these articles are not only brief but provide virtually nothing new and there is only scant mention of his contributions to quantum chemistry. Cf. Antoniotti, P.: *E. Hückel e il suo contributo alla chimica quantistica*, in: Franco Calascibetta-Eugenio Torracca (a cura di), Atti del II Convegno Nazionale di Storia e Fondamenti della Chimica, Roma, Accademia Nazionale delle Scienze detta die XL, 1988, 287–295; Frenking, G.: *100. Geburtstag von Erich Hückel*, in: Chemie in unserer Zeit 31 (1997), 27–31; Haberditzl, W.: *50 Jahre Theorie der chemischen Bindung*, in: Zeitschrift für Chemie 18 (1978), 353–359; Suchy, K.: *Erich Hückel zum Gedenken*, in: Physikalische Blätter 36 (1980), 248–249; *Erich Hückel*, in: Physics Today 33 (1980), 72–75; Beneke, K.: *Erich Hückel (09.08.1896 Berlin-16.02.1980 Marburg)*, in: *Biographien und wissenschaftliche Lebensläufe von Kolloidwissenschaftlern, deren Lebensdaten mit 1996 in Verbindung stehen. Beiträge zur Geschichte der Kolloidwissenschaften, VIII. Mitteilungen der Kolloid-Gesellschaft, 1999, 274–304*. Verlag Reinhard Knof, Nehmten 1999.

of chemistry is as rare a documentation as it is interesting. The correspondence also reveals how alluring it must have been for Hückel to try to convince this colleague of his new ideas.

The structure of this study is as follows. First, I examine and analyse Hückel's training at Göttingen, his dissertation and his various topics of research during the 1920s. Second, I analyse the emergence, development and importance of Hückel's contributions to quantum chemistry. Third, I identify Hückel's specific profile as a quantum chemist along with his characteristic style of research, his theoretical and methodological approach. In the fourth step I examine Hückel's attitude to Pauling's concept of resonance. I hope thus to arrive at a deeper understanding of Hückel's importance in the history of quantum chemistry. I close with a brief survey of a few pivotal events in Erich Hückel's professional life in National Socialist Germany.

Hückel's creative phase falls mainly within the period between 1921 and 1937. Upon completion of his doctorate in experimental physics, Hückel's research papers cover the theory of strong electrolytes, some of which he published together with his mentor Peter Debye. After 1930 he mainly published articles independently on the quantum-theoretical interpretation of the bonds of unsaturated and aromatic compounds, which constitute his *opus magnum*.[14] As mentioned above, these contributions lie fully within the new discipline of quantum chemistry that had been forming in Germany since 1927. Thus the focus of this present work is on Hückel's scientific trajectory from physics to the new hybrid discipline of quantum chemistry. I follow this career in various stages.

In Chapter 1, I start with a sketch of Hückel's family background, pointing out the enthusiasm that the three brothers Walter, Erich and Rudi had for science inspired by their father, the medical doctor and private scholar Armand Hückel. Then Erich Hückel's studies at the University of Göttingen before and after World War I are described as well as his working experience in science during the war. As a student Hückel maintained close ties with influential scientists from the Göttingen circle, including Peter Debye, David Hilbert and Max Born, who all had a major influence on his intellectual development. In the next few stages I discuss Hückel's time as a doctoral student under Debye, as a student aid for Hilbert, followed by his assistantship under Born at Göttingen and under Debye at Zurich, where he took his second academic degree (*Habilitation*) with a thesis on the theory of strong electrolytes. Then I sketch Hückel's enchantment with the new theory of quantum mechanics and his first applications of quantum mechanics to organic chemistry. There follows a description of Hückel's research stay (from mid-March to the end of June 1929) at the Niels Bohr Institute in Copenhagen. In this stimulating atmosphere he was inspired by Bohr to work on the quantum theoretical interpretation of double bonding. Finally, I portray Debye's efforts to place him at a German university. He succeeded in obtaining a one-year grant from the *Notgemeinschaft der deutschen Wissenschaft* in the fall of 1929, to conduct research at the department

---

[14] Hückel, E.: *Ein Gelehrtenleben Ernst und Satire*. Verlag Chemie, Weinheim 1975, p. 177 f.

of theoretical physics at the University of Leipzig, where Werner Heisenberg and Friedrich Hund were working. That is where Hückel posited his quantum theory of double bonding.

I discuss this and his research in quantum chemistry as a whole during the 1930s in Chapter 2. After an outline of the classical models of double bonding in stereochemistry and its controversial aspects of particular relevance to Hückel's quantum theory of double bonding, I describe the scientific collaboration between Erich and his brother Walter, as far as can be gathered from the sources. This collaboration became very close during the 1930s and played a particularly important part in Erich's research agenda in quantum chemistry. I then discuss his quantum mechanical interpretation of double bonding and its limitations. It can be shown that through quantum mechanics Erich Hückel arrived at a less intuitive and more abstract understanding of stereochemical issues than the traditional organic chemists with their visual or metaphorical representations. On this basis a comparison can be drawn between Hückel's model of double bonding and those independently developed by the Americans Linus Pauling and John Slater at the same time. It will be shown that, contrary to the Pauling and Slater model, which legitimizes the visual models of classical organic chemistry through quantum mechanics, Hückel's conceptions pointed in a certain sense to a new epistemological horizon. I then describe Hückel's efforts to publicize his first contribution to quantum chemistry in his native country, as well as the first reactions both inside and outside of Germany. I then sketch Hückel's career as an appointed lecturer of "chemical physics" at Stuttgart (1930–1937), which can be regarded as the first attempt to implement the new field of quantum chemistry in research and teaching at a German university.

At Stuttgart Hückel continued to develop his agenda in quantum chemistry. His second pioneering paper on the benzene problem was written there as well as his general quantum theory of aromaticity. Following an assessment of Hückel's research work, I show that Hückel took the approach by organic chemists in seeking a general criterium of aromaticity by which compounds could be classified as aromatic. I demonstrate furthermore how Hückel arrived theoretically and methodologically at a quantum theoretical interpretation of aromaticity. Finally, the pivotal points of his theory of aromaticity are described and commented upon. On the basis of Hückel's ideas on bond stability, I argue that Hückel offered more of a chemical explanation than a physical one for the energetic stability of aromatic ring structures. On the basis of correspondence between Hückel and Born it is shown, in addition, how Hückel arrived at the idea that the classical valence-line representation of benzene and the fused ring bonds cannot be completely understood without a more profound conceptual structure, namely, the valence structure of quantum theory. In my view, these issues indicate Hilbert's strong influence on Hückel's thought, a detailed discussion of which follows. I then go into Friedrich Hund's localization conditions, which not only support Hückel's critical considerations on Kékulé's diagram of the benzene molecule but also clearly indicate the limitations of the applicability of the valence-line diagram in organic chemistry. I then examine critically Hückel's quantum theoretical interpretation of the chemical behavior of benzene upon multiple substitutions which, together with his contribution on the benzene problem, triggered a controversy with Pauling and his pupil Georg Wheland. This is elaborated in the following two chapters. I end the second chapter with a description of the first positive responses to Hückel's quantum-theoretical treatment of chemical problems until 1934.

Before entering into the controversy between Hückel and Pauling about the benzene problem in the third chapter, I insert a historical outline of the development of the formal mathematical conventions used in spin-invariant theory, in order to illuminate the different physical interpretations reached by Hückel and Pauling and their dissimilar applications to chemistry. This gives a clearer picture not only of the scientific context in which the controversy took place, but also of the role of the quantum-mechanical theory of resonance in chemistry. There follows a detailed critical analysis of the controversy. I explain and underpin Hückel's suggested neologism "special proportion of energy" as opposed to Pauling's "resonance energy" concerning the real energetic ground state of benzene by "resonance stabilization."

In Chapters 4, 5 and 6 the attempt by Wheland and Pauling to base Christopher Ingold's general electron theory of organic chemical reactions on quantum theory by means of the "molecular orbital method" is described along with their debates with Hückel. A sketch of the criterium of magnetic aromaticity developed by Pauling then follows along with his main idea of a cyclical motion of the p_z-electrons, which Hückel shortly afterwards creatively incorporated into his theory of aromaticity. Hückel's efforts to disseminate his theory of unsaturated and aromatic bonds abroad and principally at home took place during the second half of the 1930s. The reasons for the weak reception of Hückel's ideas in Germany are presented and discussed. I close with an impartial portrayal of Hückel's scientific career under National Socialism. His personal files are taken as a basis and I also address his membership in the National Socialist Welfare Association [*Wohlfahrt*] and the main National Socialist German Workers Party.

Chapter 7 constitutes a survey of events after World War II. It contains a brief sketch of the difficulties Hückel encountered with the denazification tribunal and his reappointment as extraordinary professor of theoretical physics in Marburg. I close with the distinction awarded to him as an emeritus for his theory of aromatic bonds by the German Society of Chemists and the German Physical Society.

Chapter 1
Erich Hückel's Education and Scientific Awakening: The Path to Quantum Chemistry

A lucky star seems to have hung over the year 1896 for quantum chemistry. Three of its founders in the 20th century were born in that year: Erich Hückel, Friedrich Hund and Robert S. Mullikan.[1] Armand Arthur *Erich* Joseph Hückel was born on August 9, 1896 in Charlottenburg, on the outskirts of Berlin, as the second of three sons to Marie Hückel (1879–1947) née Maier and the lecturer and medical doctor Armand Hückel (1860–1927). He spent his first 3 years in Charlottenburg with his brother Walter, who was a year older than him, in a spatious apartment on the first story of a building on Schlüterstrasse before the arrival of their younger brother Rudi.

1.1 The Hückel Family

Erich Hückel's grandfather, Bernhard Hückel (1826–1892), came from a family of wealthy farmers originating from Hatten in Alsace (now in France). The young Bernhard had been sent to preparatory school in nearby Strasbourg, where he continued his education at the university between 1846 and 1850, first in Protestant theology and then in classical philology. His academic credentials brought him the position of steward at the court of Prince Max Roux Damiani and private tutor of his sons Nicolas and Georges. Prince Damiani was a descendant of Josephine Beauharnais, the first wife of Napoleon I. Bernhard Hückel later became head teacher at the French school in Stuttgart and then director of the city savings bank Sparkasse Stuttgart. Through his acquaintance he met the attractive and wealthy widow of a manufacturer Caroline Stälin. Her father Carl Friedrich Gärtner (1772–1850) was a celebrated doctor and botanist from Calw. Her grandfather, Joseph Gärtner (1732–1791), was likewise a botanist and a member of the Academy in Petersburg, known for his *Carpologia*. Caroline and Bernhard soon became engaged and were married in 1859.[2]

[1] Friedrich Hund was born on February 4, 1896 and Robert S. Mullikan on June 7, 1896.

[2] All information on Hückel's ancestors are taken from his autobiography. Cf. Hückel, E.: *Ein Gelehrtenleben. Ernst und Satire*. Verlag Chemie, Weinheim, 1975, Meine Vorfahren, pp. 13–28.

A. Karachalios, *Erich Hückel (1896–1980)*, Boston Studies in the Philosophy of Science 283, DOI 10.1007/978-90-481-3560-8_1,

Caroline brought along three children from her first marriage. Erich Hückel's father Armand was born from this second marriage on February 26, 1860. Hence there was a clear streak for science in the family that Armand was able to pass on to his sons. Erich reports in his autobiography:

> A fortunate fusion of the genes of the married couple, who both contributed excellent predispositions, must have contributed to the uncommon intelligence of my father and therefore his 3 sons Walter, Erich and Rudi. One may assume that my brother Walter's very exceptional aptitude to remember things was related to his grandfather Bernhard's extraordinary memory.[3]

Erich Hückel's father, the eldest child from the marriage between Bernhard and Caroline, wanted to study chemistry after completing his schooling in Stuttgart. In the opinion of his father Bernhard, however, this subject offered no professional prospects and he decided that his son choose medicine. So Armand commenced his studies in medicine and psychiatry first at Tübingen, then with various stipends at Montpellier and Paris. Afterwards he became a private lecturer and assistant at the clinic for internal medicine in Tübingen. During his university studies at Tübingen he attended lectures on various other fields in the natural sciences as well, such a mineralogy and geology, in order to broaden his scientific foundations. Hückel's father particularly enjoyed attending the chemistry lectures by Julius Lothar Meyer (1830–1895) "during the cold winter of 1879–1880."[4]

Armand Hückel met his wife at the Tübingen clinic, where she was being treated for paratyphoid. Erich Hückel describes this first meeting along with a portrait of his mother in his autobiography:

> At that time she was 18 years old, a simple, natural, cheerful, and energetic – very pretty – farmer's daughter from Nehren, a village near Tübingen. My father's affection for this "child," as girls could still be called in those days, was immediate. And my father continued to call her that, always just "child," after she became his wife.
>
> The girl too was quickly swept up in a deep affection for my father, as kind and compassionate a man as my father was.[5]

Hückel's father immediately proposed to Marie but his wish to marry this "pretty farmer's daughter" encountered violent opposition by his family, particularly by the father. Despite this meeting of kindred souls, in their day the social barrier between the two was too great. So Armand Hückel decided to remove "his" Marie from her lowly surroundings. With the acquiescence of her parents he had her enrolled in a finishing school in the vicinity of Hannover in order to complete her education and learn the manners expected of a young lady of his social standing. But the meagre earnings he made as an assistant and from the occasional consultations in his private practice did not nearly suffice to pay for it. So Armand Hückel abandoned his wish to follow an academic career and accepted a position as personal physician to

[3]Ibid., p. 16.

[4]Hückel, W.: *Memoiren* (Abschrift), UAT 311, p. 3.

[5]Hückel, E.: *Ein Gelehrtenleben*, p. 24.

Prince Henckel von Donnersmark. This step also required his leaving Tübingen for Upper Silesia (now Poland) in the region between Lublinitz and Tarnowitz, where the prince's castle Neudeck was located. He stayed there from 1891 to 1893.

In early 1892 Armand Hückel's father Bernard died and a year later "his most dearly beloved" mother Caroline followed him.[6] The quite large inheritance turned Armand into a man of means and thus he became financially independent. After his three year probation expired, he terminated his contract with the prince in order to pursue private research. In 1894 he married his Marie in Döhren near Hannover. Erich Hückel recounts in his autobiography that, fearing the gossip among his colleagues about his wife's "mean origins," his father was averse to returning to provincial Tübingen.[7]

Meanwhile Erich Hückel's father had carefully prepared for his new life as a private scholar. Before the wedding he rented the above-mentioned apartment on Schlüterstrasse in Charlottenburg near Berlin. His first project was to identify the pathogen of small pox. After roughly four years of experimental investigations he published a book under the title "*The Vaccine Corpuscles*" in which he disproved claims that the vaccine was contagious. But he was not able to find the pathogen itself. He came to the conclusion that it was too small to be detected by microscope.[8]

Hückel's father enjoyed his research in Berlin but the local university snubbed his advances to become academically active. Hückel writes about this in his autobiography:

> The University of Berlin declined his petition to teach as a private lecturer on the justification that his *Habilitation* thesis from Tübingen did not meet the requirements valid in Berlin and that the same applied to his scientific researches. This rejection hurt my father deeply so he decided to leave Berlin and move to a smaller town. The study on vaccine corpuscles had, incidentally, not been submitted to the medical faculty in Berlin.[9]

Under such circumstances Hückel's father decided to move his family near another university. He was moreover of the opinion that such a large city as Berlin was not particularly suitable for the healthy development of his children. The new home would have to be a smaller academic town. He wanted his children to grow up not only in closer touch with nature but also in a culturally and intellectually inspiring milieu. Göttingen was thus an ideal choice. Its university, the Georgia Augusta, enjoyed a high reputation primarily in mathematics and the natural sciences.

He bought a family home in the eastern end of this small university town at Friedländerweg no. 46. On May 25, 1899 the whole family moved into their new neighborhood and the three boys grew up in the fine new house. Their household included a governess, a cook and a maid. The lifestyle in the Hückel home "had a

---

[6]Ibid., p. 26.

[7]Ibid., p. 26

[8]Cf. Hückel, A.: *Die Vaccinekörperchen. Nach Untersuchungen an der geimpften Hornhaut des Kaninchens. Ein experimenteller und kritischer Beitrag zur Frage nach dem Contagium der Vaccine*. Gustav Fischer, Jena, 1898.

[9]Hückel, E.: *Ein Gelehrtenleben*, pp. 27–28.

Frenchness about it, like my grandfather's family. Our home life was quite 'noble' but simple."[10]

1.2 The First Years in Göttingen: Childhood, Elementary and Preparatory Schooling

The property in Göttingen also included two gardens, along with a large field that Hückel's father had also purchased. So there was plenty of nature to see. The observation of nature was a primary object of their father Armand, who took them on various shorter walks as well as on the occasional excursion in the vicinity of Göttingen. On these outings Armand taught his children the names of the various plants and mushrooms they encountered along the way. They became familiar with various butterfly species and from their own beetle and butterfly collections they learned about the metamorphosis of insects. They also had a terrarium with lizards at home and a large collection of minerals and fossils.

It was sheer good fortune for the boys that their father abandoned his life's dream of becoming a university professor and decided to devote his time and energy to their upbringing. The three brothers were able to profit fully from their father's vibrant interest in science. In 1904 he bought a telescope with which his sons could observe the lunar planets, Saturn's rings and the more remote binary stars. In the basement Armand set up a chemistry laboratory equipped with a ventilator, and a room on the first story was furnished as a workshop and physics laboratory. In the workshop he built salt-water acquariums for various species of fish, shellfish, crabs and marine flora. Erich learned there how to handle various tools, like the carpenter's bench and lathe. His father built various physical instruments himself and, as Erich later described, "models to demonstrate Maxwellian electrodynamics. He even replicated Heinrich Hertz's experiments on electromagnetic waves. He also constructed a large electrostatic machine with Leyden jars that produced long, bright, loudly crackling sparks. He also made a large inductor."[11]

This laboratory was used not only for experimentation but also for the serious study of chemistry. With the aid of their father the Hückel sons read early on Wilhelm Ostwald's *Die Schule der Chemie* as an introduction to the basic concepts of chemistry. Erich's brother Walter describes in his memoirs the first chemistry experiments in their private laboratory and the practical manuals they consulted:

> For Christmas 1907 we received [at the ages of 12, 11 and 8] a few chemicals and simple glass apparatus as gifts. During the Christmas holidays my father conducted with my brother Erich and me our first chemical experiments, which he initially thought up himself. Heumann's work *Chemische Experimente*, written for lectures on anorganic chemistry (experimental chemistry), also provided instructions for many nice experiments. The quantitative aspect of chemical processes was an important factor in the choice of which

[10]Ibid., p. 34.

[11]Ibid., p. 47.

experiments to conduct. Later, experiments from Stöckhardt's *Schule der Chemie* and from Scheidt and Lowenhardt were also performed. At the same time, an analytic balance that my father had bought was used to practice weighing. These quantative experiments were conducted only later, of course, around 1910. About the same time, perhaps a little earlier, there was the systematic course of instruction on the reactions of the commonest salts, which we followed according to a little book by Fichter. [...] That's because my father started going through Ostwald's *Schule der Chemie* at the beginning of 1909. [...] The individual sections were read out, whereby we found the dialog form particularly stimulating; we were able to add our own questions to liven up the dialog in the book. We caught up on the feasible experiments we had not yet performed. [...]

The pedagogically quite exceptionally written *Schule der Chemie* by Ostwald was of decisive importance as a foundation for my scientific development. I became acquainted with anorganic chemistry at its latest stage and thus did not need to waste time on correcting obsolete ideas. [...] Nor does the importance of physical processes in chemistry get shortchanged. [...] A major advantage of this book was its heavy emphasis on the theory of ions, which immediately brought the theory of electricity into the correct light in chemistry as well.[12]

The Hückels' educational program incorporated music as well as literature. Erich and Walter took piano lessons for a long time, while Rudi learned to play the cello. Their father carefully chose their reading material. It consisted primarily of juvenile literature but also included ballads by Schiller and poetry by Goethe.

At the age of six, Erich started attending an elementary school in 1902 for three years. Afterwards from 1905 to 1914 he went to the royal preparatory school, the *Gymnasium* at Theaterplatz in his home town, which he completed with a diploma. Looking back on these school days in Göttingen, Erich wrote in his autobiography:

I value what we learned and experienced with my father very much more highly than anything we were assigned or not assigned during those 9 years of imprisonment in the preparatory school.[13]

Erich's brother Walter gives a more detailed report of this period in his memoirs:

What I am grateful to my father for during our preparatory schooling is not the help he gave us in mastering the homework. My father transmitted to my brothers and me, as before and in continuation of his former tutoring, a broad theoretical and practical foundation in the area of, indeed, all the natural sciences, not in a rigid curriculum but half in play.[14]

So it is not surprising that after passing their school leaving examinations the three sons were fully qualified to commence studies in science at their local university. Walter started "exuberantly and enthusiastically" on April 16, 1913 what had been "his choice of study for years." Rudi chose medicine and became a pathologist.[15] As a schoolboy Erich too had expressed a wish to study medicine. Erich explains in his authobiography in this regard: "My father advised against medicine; that so very responsible profession would overwhelm my sensitive nature

[12] Hückel, W.: *Memoiren*, pp. 20–22.

[13] Hückel, E.: *Ein Gelehrtenleben*, p. 43.

[14] Hückel, W.: *Memoiren*, p. 18.

[15] Hückel, E.: *Ein Gelehrtenleben*, p. 48.

and my conscientiousness."[16] Although Erich followed his father's advice, his interest in medical problems remained with him throughout his life. His later reading also included books on psychology.

1.3 Beginnings at the University of Göttingen and the War Years

Thus Erich's decision to study physics was reached only after his graduation from school. At Easter 1914 shortly before the outbreak of World War I, he enrolled in "physics in particular" and "mathematics in general" at the mathematics and science department of the philosophical faculty of the Georg August University of Göttingen. In the first semester Erich attended "far too many lectures," as he puts it in his autobiography, for which he was "not scientifically mature enough."[17]

Because Göttingen had a particularly strong curriculum in the natural sciences, Erich tried to follow his brother Walter's example and obtain as broad a foundation as possible from his very first semester on. His courses were not limited to the required lectures for aspiring physicists and mathematicians. He also attended the lecture on organic chemistry by the Nobel laureate Otto Wallach (1847–1931).[18] It is very probable that Erich took this course with his brother Walter, because soon afterwards Wallach personally intervened in Walter's later course of study.[19] In the summer term of 1914 he assigned Walter the task of carrying out a quantitative analysis in the laboratory. When Walter quickly grasped the problem and obtained very accurate results, Wallach's praise acted as a great boost to his self-confidence.[20]

From Erich's curriculum vitae we gather that he also attended courses on inorganic chemistry. At that time, Professor Arthur Kötz directed the inorganic department of the Institut für Chemie. Walter describes the mood there in his memoirs as follows:

> Prof. Kötz was a very kindly teacher who not only tried to initiate his young students, be they chemists or scientists of any other description, into the science of chemistry, but was interested in getting to know the students personally as well. As the assistants Mähnert and Wienhaus were likewise very congenial, a truly casual and friendly atmosphere reigned in the inorganic auditorium; beer evenings in the laboratory were organized as well as excursions and parties.[21]

---

[16] Ibid., p. 48

[17] Ibid., p. 69.

[18] Cf. Erich Hückels Lebenslauf in: Hückel, E.: *Dissertation*, Göttingen, 1921.

[19] Concerning Wallach's lecture Walter recalls in his memoirs: "I attended the great lecture course on experimental chemistry, starting with organic chemistry, by Privy Councillor Wallach, 5 hours in the summer and 6 hours in the winter. It was the last but one lecture he offered before retiring." Hückel, W.: *Memoiren*, p. 50. Wallach became emeritus in 1915.

[20] Hückel, W.: *Memoiren*, p. 46. Cf. Hanack, M.: *Walter Hückel (1895–1973)*, in: CB 113 (1980), I–XXVIII.

[21] Hückel, W.: *Memoiren*, p. 43.

Following their father's wishes Erich and his brother Walter planned to leave the Georgia Augusta for a few semesters to study at another university. Freiburg was under consideration. But the war put an end to this plan and both brothers spent their entire time as students at Göttingen.[22] Erich remembered the mood at the beginning of August at the outbreak of war:

> The German nation was swept up in a wave of patriotism in heady anticipation of victory. [...]
>
> Kaiser Wilhelm II gave the clairvoyant pronouncement: "When the foliage falls from the trees, you will be home again."
>
> The youth, mostly students and high-school boys, came in droves to enlist as volunteers. Walter and I did so as well, I more because everyone else was doing so and I did not think I could exclude myself. But we were fortunate enough to be turned away as too frail.[23]

So Erich continued his studies at Göttingen until 1916. Being only partially fit for military service, he was first assigned as a scientific aid under Ludwig Prandtl, professor of applied mechanics at the "Model Testing Station in Aerodynamics" in Göttingen.[24] There followed until the end of the war his deployment as an assistant at the "Sea-Plane Testing Headquarters" in Warnemünde. The Model Testing Station founded by Prandtl was at that time laying the corner-stone of its future fame. During the war Prandtl and his coworkers were fully occupied with developing the first successful procedures for calculating the aerodynamics of airplanes.[25] Prandtl knew Erich from one of his practical courses and was happy to employ him at his station. Erich found this position convenient not just because of the war. He apparently enjoyed it as well. Hückel outlined the station's goals and his own tasks as follows:

> The task of the Model Testing Station was to determine the forces exerted on an airplane in flight by measuring them on a model and then calculating the forces on the airplane itself from the results obtained from the model. This is what I had to do – mainly under the direction of the engineer Dr. Max Munk, a department head of this station: take measurements in the "wind tunnel" in which the models were suspended in an air current, as well as perform the calculations. Mr. Munk was considerably older than I but we were soon on very friendly terms. [...]
>
> I then mainly had to review the calculations and check them for accuracy. Because the calculations always followed the same procedure, it was relatively easy for me to check them. I usually only had to glance through the series of numbers to find the errors simply because the numbers did not fit within the series.[26]

---

[22]Hückel, W.: *Memoiren*, pp. 41–42. Cf. Hückel, E.: *Ein Gelehrtenleben*, p. 71.

[23]Hückel, E.: *Ein Gelehrtenleben*, p. 69.

[24]This institute was later expanded and renamed "Aerodynamic Testing Station," because the research expanded beyond model testing.

[25]Cf. Rotta, J. C.: *Die Aerodynamische Versuchsanstalt in Göttigen, ein Werk Ludwig Prandtls. Ihre Geschichte von den Anfängen bis 1925*. Vandenhoeck & Ruprecht, Göttingen, 1990.

[26]Hückel, E.: *Ein Gelehrtenleben*, p. 71. Munk was one of Prandtl's leading theoretical collaborators during this phase. Shortly after World War I, Munk emigrated to the U.S. and was a major contributor to the development of aerodynamics there. Cf. Anderson, J. D.: *A History of*

In 1918 Erich was drafted in the army. Munk, who had left the Model Testing Station and was working as an engineer at the Sea-Plane Testing Headquarters, sent for Erich to serve as his assistant there. This exempted him of his army service and he was ordered to go to Kiel as a "sea-plane sailor." Erich described his period of "basic training as a soldier" in Kiel as "the most mind-numbing time" of his life. Afterwards he was transferred to the "Sea-Plane Testing Headquarters" in Warnemünde by the Baltic.[27] His tasks there consisted in "taking in-flight measurements of and inside mostly single-engine monoplanes and then evaluating them."[28] The results of some of these aerodynamic investigations appeared 1917/18 in *Technische Berichte der Flugmeisterei*.[29] This was Erich Hückel's debut in science.[30]

1.4 Continued Studies at Göttingen

With the return of peace Erich continued his university studies at Göttingen. In his autobiography he recalled that during this period the lecture material was much less demanding and incorporated fewer exercises.[31] He learned theoretical physics from his own private study rather than from any courses he attended during that time. As regards his mathematical training, he tells us that he took a standard course on differential calculus from which he learned practically nothing.[32] So his mathematical skills remained limited and he had to teach himself the more advanced mathematics later on as an assistant and professor. Despite these shortcomings in the curriculum, Peter Debye's lecture during the winter term 1919/20 on "Atomic structure

Aerodynamics and its Impact on Flying Machines. Cambridge University Press, Cambridge, 1997, pp. 289–292.

[27] Trischler, H.: *Luft- und Raumfahrtforschung in Deutschland 1900–1970: Politische Geschichte einer Wissenschaft*. Frankfurt am Main, Campus, 1992.

[28] Hückel, E.: *Ein Gelehrtenleben*, p. 74.

[29] Munk, M., Hückel, E.: *Systematische Messungen an Flügelprofilen*, in: Mitteilungen der Göttinger Modellversuchsanstalt für Aerodynamik, Nr. 2, erschienen in Technische Berichte (1917), pp. 148–163; *Weitere Untersuchungen von Flügelprofilen*, in: Mitt. d. MVA, Nr. 6, erschienen in Technische Berichte (1917), pp. 204–218; *Weitere Göttinger Flügeluntersuchungen*, in: Mitt. d. MVA, Nr. 17, erschienen in Technische Berichte (1918), pp. 407–450; *Der Profilwiderstand von Tragflügeln. Eine Zusammenfassung der bisherigen Göttinger Flügelprofilmessungen*, in: Mitt. d. MVA, Nr. 18, erschienen in Technische Berichte (1918), pp. 451–461. After World War I ended Hückel was decorated with a medal for distinguished wartime service together with fellow coworkers by the (democratic) Prussian government. Cf. Rotta, J. C.: *Die Aerodynamische Versuchsanstalt in Göttigen, ein Werk Ludwig Prandtls. Ihre Geschichte von den Anfängen bis 1925*. Vandenhoeck & Ruprecht, Göttingen, 1990, p. 197.

[30] Cf. Kragh, H.: *Before quantum chemistry: Erich Hückel and the physics-chemistry interface*, in: CENTAURUS 43 (2001), 1–16.

[31] Hückel, E.: *Ein Gelehrtenleben*, p. 77.

[32] Ibid.

and quantum of action" was an important impulse for Hückel. Debye covered general problems of quantum theory in great depth, and thus Erich Hückel gained his first impression of the main concepts and methods of the "older quantum theory" as it is now often called in retrospect. Together with Hans Falkenhagen and Herbert Ebert he carefully reworked the course material and the result was a publishable manuscript.[33] As Hückel mentions in his autobiography, Debye thought it untimely to publish it, however, because Arnold Sommerfeld's book on a related topic, *Atomic Structure and Spectral Lines*, had just appeared shortly before.[34] Debye's lecture fascinated the young Hückel. He decided to visit Debye in his private apartment to ask him to take him on as one of his pupils, because "he had the intention of becoming a scientist."[35] Debye's initial reaction was cautious. But he later accepted him as one of his graduate students, assigning him as a thesis the experimental topic: "Scattering of Röntgen rays by anisotropic fluids."

1.5 Hückel's Doctorate in Experimental Physics

Certain substances of a specific chemical structure melt at a specific temperature into an opaque "crystalline" liquid that is optically anisotropic. Such liquids exhibit optical properties that are otherwise only found in solid crystals. If one continues to heat them beyond their melting point to a given temperature (called their "clarification point") the melted liquid becomes a clear isotropic liquid. The Austrian botanist Friedrich Reinitzer was the first to observe this phenomenon of liquid crystalline substances in 1888. The physicist and crystallographer Otto Lehmann proposed the first explanation of this state in 1889.[36] A few years afterwards Lehmann divided these substances into two classes, which he termed "fluid and liquid crystals" ["fliessende bzw. flüssige Kristalle"].[37] This discovery of liquid crystals was met with skepticism among many of Lehmann's colleagues, particularly among physical chemists. Notably Walter Nernst and the physico-chemist at Göttingen Gustav

---

[33] In this regard Hückel recalls in his autobiography: "The most fruitful basis I obtained during that time of study was from an elaboration of Debye's lecture on 'Atomic structure and quantum of action' in 1920. Together with Hans Falkenhagen and Hermann Ebert I reworked my notes so thoroughly that it could have appeared as a book. It never came to that because Arnold Sommerfeld's work *Atomic Structure and Spectral Lines* had just appeared. The topic of this work covered a more limited area of quantum theory but was very detailed in that area. Debye's lecture covered a broader scope of general problems in quantum theory but was much less detailed in the special area of spectral theory. Even so, it seemed untimely to Debye to release another book with a similar aim right after Sommerfeld's." Ibid., pp. 77–78.

[34] A copy of it exists entitled *Über den Bau der Atome und Molekuele*, in: NSUB, Cod. Ms. 1993. 27.

[35] Hückel, E.: *Ein Gelehrtenleben*, p. 78.

[36] Cf. Knoll, P.: *Otto Lehmann, Erforscher der flüssigen Kristalle*. P. M. Knoll, Ettlingen, 1988.

[37] Lehmann, O.: *Flüssige Kristalle, sowie Plastizität von Kristallen im Allgemeinen, Molekulare Umlagerungen und Aggregatzustandsänderungen*. Leipzig, 1904.

Tammann were among the doubters. They believed the observed crystalline properties had to be attributable to emulsion formation owing to contaminants, impurities and foaming. A controversy arose that persisted for 30 years.[38]

By 1904 roughly 35 crystalline liquid substances were known,[39] but no correlation could be drawn between the occurrence of this liquid-crystal state[40] and molecular structure. This was achieved by the organic chemist Daniel Vorländer. Over the course of the next three years he found that such a state depended on the molecular composition of liquid crystals and demonstrated a common linear molecular structure among its various forms.[41] In 1908 Vorländer published a summary of the latest chemical research in the field of liquid crystals.[42] As Vorländer's chronicler Conrad Weyngard later explained, Vorländer made it his life's work to develop the theory of liquid crystals.[43] In 1908 he also launched an article[44] against the emulsion hypothesis of liquid crystals that Tammann and Walter Nernst were still upholding. So the debate raged on. Vorländer demonstrated the existence of anisotropic liquid phases.

Thus Hückel's dissertation addressed a very controversial issue. His goal was to find out "whether or not space lattice structures occur in substances that, according to O. Lehmann, form fluid or liquid crystals."[45] So it essentially involved finding out whether such melts possess true crystalline structures.

The method Hückel used to detect such space lattices was the one that Debye had developed together with the experimental physicist Paul Scherrer. They had used the interference of x rays to determine the atomic structure of crystals.[46] Scherrer, a private lecturer at Göttingen until 1920, also assisted Hückel in carrying out the experiments.

From among the group of "liquid" crystals, Hückel examined three organic substances that Vorländer had given him: para-azoxy anisole, para-azoxy phenetole and

---

[38] Cf. Knoll, P.: *Otto Lehmann, Erforscher der flüssigen Kristalle.* P. M. Knoll, Ettlingen, 1988.

[39] These compounds included phenol ether and azo benzoic ester.

[40] Today we refer to a "mesophase." Cf. Atkins, P. W.: *Einführung in die Physikalische Chemie.* VCH, Weinheim, 1993.

[41] Vorländer, D.: *Ueber krystallinisch-flüssige Substanzen*, in: B 39 (1906), 803–810; *Einfluß der molekularen Gestalt auf den krystallinisch-flüssigen Zustand*, in: B 40 (1907), 1970–1972.

[42] Vorländer, D.: *Krystallinisch-flüssige Substanzen.* F. Enke, Stuttgart, 1908.

[43] Weyngard, C.: *Daniel Vorländer*, in: B 76 (1943), 41–58.

[44] Vorländer, D., Kasten, W.: *Ueber durchsichtig klare krystallinische Flüssigkeiten*, in: B 41 (1908), 2033–2052.

[45] Hückel, E.: *Zerstreuung von Röntgenstrahlen durch anisotrope Flüssigkeiten.* Dissertation, Göttingen, 1921, p. 4.

[46] Debye, P., Scherrer, P.: *Interferenzen an regellos orientierten Teilchen im Röntgenlicht. I*, in: PZ 17(1916), 277–283; *Interferenzen an regellos orientierten Teilchen im Röntgenlicht. II*, in: PZ 18 (1917), 291–301.

dialisalasin. From among the "fluid" crystals he analysed cholesteric propionate and cholesteric benzoate from samples that his brother Walter had produced.[47]

Hückel summarized the results of his experimental research as follows: "So-called liquid and fluid crystals have no space lattice structure. If there are any space lattices, they occur at best in negligible quantities in crystals suspended in the liquid."[48]

Hückel was working with a liquid that consists of irratically arranged molecules rather than with a crystal powder. So he posed the additional question: "was the spatial distribution of intensity of the scattered radiation determined by a regular arrangement of atoms in the molecules?" Hückel's answer followed the view that Debye had expressed in a lecture at the 86th convention of German Scientists and Medical Doctors in September 1920: "the main interference ring observed in liquids is not determined by the internal molecular structure." Hückel supports this claim in his dissertation as follows:

> It can be shown that the cause of the [main interference ring] is the finite distances at which molecules can approach each other and consequently its angular aperture is substantially determined by the relation between the wavelength used and molecular diameter. Thus it does not yield any characteristics of internal molecular structure but rather yields a determination of molecular diameter.[49]

Reading between the lines one can see that Debye had alerted young Hückel to the connection between chemical composition (molecular structure) and physical properties. I must point out here that this dissertation showed Hückel how to transgress disciplinary boundaries in organic chemistry in order to apply new experimental methods and conceptual approaches.

Hückel distanced himself from Vorländer's foregoing views about the cause of such a liquid-crystal state with its physical properties of optical anisotropy. In Lehmann's footsteps Vorländer had contended that this state with its associated properties was not caused by a kind of aggregation of the molecules but by their chemical constitution (molecular structure).[50] Hückel argued, on the contrary, that another explanation had to sought for the occurrence of the properties of such substances because, as he had already proven, liquid and fluid crystals exhibited no space lattices. One was essentially dealing with liquids, hence, Hückel concluded, the focus must lie on the properties of the molecules.

According to the conceptions of the time, all molecular forces were of an electromagnetic nature. Thus Hückel asserted in his dissertation: "the behavior of the fluids in question [must] also be governed by electric or magnetic properties of

---

[47]Hückel, E.: *Dissertation*, § 3 Das Versuchmaterial, p. 21. It has to be pointed out that the compositional formulas (structural formulas) of the substances his brother had produced were not yet fully known at the time.

[48]Ibid., p. 25.

[49]Ibid., pp. 10–11.

[50]Vorländer, D.: *Einfluß der molekularen Gestalt auf den krystallinisch-flüssigen Zustand*, in: B 40 (1907), 1970–1972.

molecules."[51] A few years earlier Max Born had followed this course in first attempting to postulate a quantitative theory of anisotropic fluids based purely on charges.[52] Debye pointed out Born's ideas in a letter to his doctoral student, but added a caution:

> First about your conversation with Tammann. I don't intend to criticize his opinion here and just want to say that an honest scientist cannot do otherwise than publish *without the least constraint* results he deems correct, whether they be of an instructive or a personal nature. I for my part am not completely convinced that Born's theory is correct. The liquid crystals exhibit such peculiar properties from a magnetic point of view that one shouldn't consider just electric dipoles but magnetic ones as well. Furthermore, do be careful in determining the moments. Liquids with dipole moments as large as Born wishes would have to associate quite heftily. Thus nothing can be concluded about the moment from a measurement of the d. c.[53] of the pure substance. One would have to use very dilute solutions indeed. Measuring the d. c. would, of course, be useful in any case, but do be careful with your conclusions![54]

We learn from this that Hückel discussed his dissertation with the celebrated physical chemist in Göttingen Tammann, apparently also debating Born's theory of anisotropic fluids with him. In order to arrive at a complete theory, Debye advised his graduate student, both the electric and magnetic forces had to be taken into account. He also emphasized the necessity of additional measurements of the dielectric constant at various concentrations to determine the dipole moments. Hückel's dissertation does not incorporate these suggested measurements but it draws on Debye's advice in its concluding words:

> Should it be the case, as would be expected, that the electric forces played an essential part besides the magnetic forces, this relation between the moment and the clarification temperature could not be satisfied. Debye's theory of a temperature dependence of the dielectric constants does allow a determination of the electric moments of the molecules. This will be attempted in another paper in order thus to obtain a touchstone for Born's theory.[55]

Hückel never did publish the promised experimental paper. Perhaps the hard working conditions at the laboratory were the reason. Under postwar conditions laboratory work was extremely difficult and unpleasant. Hückel describes them humorously and with subtle irony in his autobiography: "Advantage: I learned to tap dance, to avoid getting cold feet – the institute was not heated. Disadvantages: power failures, intermittent delays in gas deliveries, shortage of supplies like rubber, petroleum, etc. A clown could perform a grotesque about this."[56] According to

---

[51] Hückel, E.: *Dissertation*, p. 26.

[52] Born, M.: *Über die Maxwellsche Beziehung zwischen Brechungsindex und Dielektrizitätskonstante und über eine Methode zur Bestimmung der Ionenladung von Kristallen*, in: S. B. Preuß. Akad. Wiss. Berlin 1918, pp. 604–613; *Über die ultraroten Eigenschwingungen zweiatomiger Kristalle*, in: PZ 19 (1918), 539–548.

[53] This refers to the dielectric constant. This material property provides information about the behavior of a large number of molecules in an electric field.

[54] Debye an Hückel, Zürich, 26. Okt. 1920, MPG-Archiv, III. Abt., Rep. 19, NL Debye.

[55] Hückel, E.: *Dissertation*, p. 27.

[56] Hückel, E.: *Ein Gelehrtenleben*, p. 79.

Annemarie Hückel's brief biographical essay about her husband, these unusual conditions caused him to develop an aversion for experimental research. His subsequent focus became increasingly theoretical.[57]

Despite such external obstacles Hückel managed to complete his dissertation and he passed his examination on January 26, 1921.[58] For the oral part, the geophysicist Emil Wiechert examined him on theoretical physics, the mathematician David Hilbert on mathematics and Ludwig Prandtl on experimental physics. His doctoral advisor Debye was not among the examiners because he had left for Zurich in spring 1920 to take up an appointment at the Federal Polytechnic as professor of experimental physics and director of the physical laboratory.[59] Nevertheless Debye did not neglect his responsibilities towards his doctoral student. He read Hückel's dissertation carefully and forwarded his report promptly at the beginning of January 1921 to "Privy Councillor Wiechert."[60] Wiechert may have consulted Debye's report when he evaluated Hückel's dissertation as follows: "The experiments have been performed skillfully and with great care and obtain a result that is of very good scientific value."[61] Looking back on his dissertation Hückel himself very modestly thought, "leaving aside the success in surmounting the external obstacles, the experimental achievement itself was nothing spectacular."[62] Hückel is probably being overly severe about his accomplishment. One only has to look at the chapter on "The apparatus and the experimental procedure" to make one's own critical assessment of his dissertation. Hückel's careful work becomes immediately apparent both in his construction of the necessary apparatus and in his performance of the experiments with monochromatic x rays. Although there is no documentation, we can presume that the results of his experiments were not left unnoticed by the combatants in the controversy between chemists and physical chemists mentioned earlier.

The minutes of the oral examination on January 26, 1921 reveal that Hückel was familiar with a relatively broad range of modern topics of the day. Wiechert asked him, for instance, about the basic concepts of relativity theory, the Lorentz transformations and Minkowski's space-time theory. Hilbert asked him about the integral theory of normal differential equations by Hamilton and Jacobi and the corresponding applications to quantum theory. Prandtl asked him questions that related directly to experimental practice, such as the mechanics of measuring instruments, taking elasticity measurements and the modulus of elasticity. As already mentioned in the previous chapter, Hückel had maintained contacts with Prandtl before and during

---

[57]Hückel, A.: *Wer ist's? Erich Hückel*, in: Nachrichten aus Chemie und Technik 13 (1965), 382–383.

[58]A summary of his dissertation appeared under the title "*Zerstreuung von Röntgenstrahlen durch anisotrope Flüssigkeiten*", in: PZ 22 (1921), 561–563.

[59]Cf. Davies, M.: *Peter J. Debye (1884–1966)*, in: Biographical Memoirs of Fellows of The Royal Society; The Royal Society, London, vol. 16 (1970), 174–132.

[60]Debye to Hückel, Zürich, 7. Jan. 1921, MPG-Archiv, III. Abt., Rep. 19, NL Debye.

[61]UAG, Erich Hückel's doctoral procedure, in: Promotionen der Phil. Fak. HV, Nr. 24.

[62]Hückel, E.: *Ein Gelehrtenleben*, p. 80.

the war. So Prandtl's comment in his evaluation may be of interest: "The material was, considering it was learned before the war, not always secure. The overall result was good." Hilbert also judged Hückel's qualifications as "good." Wiechert's assessment of the doctoral candidate's performance was, by contrast, "excellent."[63]

Altogether, Hückel's thesis and oral examination earned the predicate "very good."[64]

1.6 Excursion: First Collaboration with His Brother Walter

Hückel's dissertation in experimental physics treated a live topic situated in the fuzzy area between solid state chemistry and molecular physics. Besides being conversant in physics and mathematics, he was also at home in chemistry. This was thanks to his brother Walter, who enabled Erich to expand his expertise to include the concepts, methods and approaches in chemistry. Erich was just a year behind his brother when he passed his doctoral examination. Walter had defended his doctorate on the chemistry of Blanc's reaction on July 7, 1920 under the supervision of the later Nobel laureate Adolf Windaus. His was examined in the subjects of chemistry, physical chemistry and physics. Walter then was employed as Windaus's assistant while working toward his next goal, the *Habilitation* degree, the qualification for academic teaching.[65] A talk held in early March 1921 at Göttingen by the physicist Walter Kossel on his own contribution to the development of the electron theory of valence turned out to be a pivotal event in Walter's development.[66] Kossel's "scholarly persona" impressed Walter. Moreover, as he later recalled: "it was after that talk that I realized how inadequate this theory actually was in the field of organic chemistry. It robbed me of my sleep that night from the 9th to the 10th of March 1921, and my paper on 'Contributions to the conception of homopolar bonding of atoms' was composed that night."[67]

Walter's paper was a critical derivative of Kossel's concept of homopolar vs. heteropolar bonds. It proposed to contribute toward the understanding of bond transitions from heteropolar to homopolar "bond types." He also specially treated a series of homopolar bonds "with similar electron arrangements," what are known as isoelectronic bonds.[68]

---

63 UAG, Promotionsvorgang von Erich Hückel, in: Promotionen der Phil. Fak. HV, Nr. 24.

64 Ibid. According to the minutes, the oral examination started at 6 o'clock sharp in the morning (!).

65 Neidlein, R.: *Walter Hückel (1895–1973)*, in: CB 113 (1980), I–V.

66 Hückel, W.: *Memoiren*, p. 201.

67 Ibid., p. 201.

68 Hückel, W.: *Beiträge zur Auffassung der Homöopolaren Atombildung*, in: Z. Elektrochem. 27 (1921), 305–309.

A later chapter will discuss Walter's ideas about isoelectronic bonds and how they influenced Erich's later quantum theory of double bonding.[69] I only note here that Erich acted as his consultant throughout. Walter closed his paper in explicit appreciation of his brother Erich's "numerous suggestions."[70] It was due to his brother's paper that Erich studied the model Bohr and Debye had proposed of the hydrogen atom as well as the cubic model of the chlorine molecule by Born and Landé along with the latest developments of valence theory in Germany up to that time in 1921.[71]

1.7 Intermediate Stops

1.7.1 David Hilbert's Auxiliary Assistant

After passing his doctoral examination Hückel decided to stay in academia and not look for a job as a commercial physicist or high-school teacher. Debye had already promised to engage him right after the conferral of his degree as one of his assistants at the Federal Polytechnic (ETH) in Zurich. Swiss regulations on the employment of foreigners, which were rigid already in those days, thwarted this plan initially. Despite Debye's many efforts, Hückel had to wait. In a letter from Zurich Debye wrote him: "As concerns the assistantship, a few things still have to be dealt with that will probably take another fortnight. It is not yet certain whether this summer term will work, but quite probably, so please be patient for a little while longer."[72]

At this moment of need Hilbert unexpectedly appeared as *deus ex machina* and offered Hückel a one year position as auxiliary assistant. Hilbert's chair included one assistantship and one auxiliary assistantship. At that time the mathematician Paul Bernays filled the assistantship. The auxiliary assistantship had been placed at Hilbert's disposal to help satisfy his ambition to throw bridges between mathematics and physics. This additional subordinate position was cut out for theoretical physicists of the Sommerfeld school, because whenever it became vacant Hilbert regularly sent for one of Sommerfeld's pupils. It was jokingly called "Hilbert's private tutor in physics,"[73] because with their assistance he was able to keep up to date on the latest developments in theoretical physics.[74] Paul Ewald, Alfred Landé

[69]See subsection 2.1.4.

[70]Hückel, W.: *Beiträge zur Auffassung der Homöopolaren Atombildung*, p. 309.

[71]About the development of valency theory in Germany until 1921 cf. Buttker, K.: *Widersprüche der Entwicklung – Entwicklung der Widersprüche. Die Herausbildung der Quantenchemie im Blickfeld philosophischer Analyse*. VEB Deutscher Verlag der Wissenschaften, Berlin, 1988, pp. 12–29.

[72]Debye to Hückel, Zürich, 7. Jan. 1921, MPG-Archiv, III. Abt., Rep. 19, NL Debye.

[73]Hückel, E.: *Ein Gelehrtenleben*, p. 81.

[74]Cf. Reid, C.: *Hilbert*. Springer-Verlag, Berlin, 1972, Chap. XVI Physics, pp. 125–147.

and Adolf Kratzer had been the initiators of this arrangement since 1912.[75] At the beginning of the summer term in 1921 the position had just become vacant because the recent auxiliary assistant Kratzer had received an appointment as full professor of theoretical physics at the University of Münster. So Hückel accepted Hilbert's offer to succeed Kratzer for 1 year at the mathematics department of the University of Göttingen. It remained a puzzle to Hückel, however, why Hilbert had offered him the position. Hilbert was mainly interested in theoretical physics and Erich had taken his doctorate in experimental physics. Hückel surmises in his autobiography:

> Maybe he liked the talks I had delivered at the physics colloquium and at the seminar on the structure of matter, at which he had been present. It is true that I tended to prepare my talks particularly carefully, spoke freely (relying only on a few brief notes) and even practiced in front of a blackboard at home to make sure I kept within the time I had set for myself. [...] On the other hand, I had a very heated argument with Hilbert during the discussion after one of my talks. It concerned one of Debye's theoretical papers on the repulsive forces that generally occur when atoms approach each other very closely. Some untenable ideas were developed on the basis of considerations in this paper. Hilbert realized this and raised objections against the theory I had just reported on. I defended Debye's (incorrect) theoretical argument with extraordinary steadfastness, because I did not see the validity of Hilbert's objections. Such an impassioned professional debate had surely never before taken place *coram publico* between a professor and a student – who was not even right.[76]

I must add that Hückel regularly attended Hilbert's lectures. Between the winter term 1916/17 and the summer of 1921 he was registered in the following courses: "The foundations of physics II"; "Space and time"; "Advanced mechanics and recent gravitational theories"; and "On geometry and physics". One of Hilbert's lectures during the winter term 1919/20 on "Conceptual methods in the exact sciences" was broadly conceived for students from the various faculties.[77] Hilbert's explanations of his conceptions of theoretical science probably had an influence on Hückel, as he remarked in his autobiography: "The details of what he presented in this lecture have escaped my memory but it surely did not fail to affect my conception of these sciences."[78]

Hilbert's lecture presented in a popularly comprehensible manner his own considerations and ideas not just about mathematics and mathematical thought, but also about physics and its conceptual systems. By means of a parallel conceptual analysis of mathematics and physics he underscored their mutual relations as well as the

---

[75]Cf. Eckert, M.: *Die Atomphysiker. Eine Geschichte der theoretischen Physik am Beispiel der Sommerfeldschule*. Vieweg, Braunschweig/Wiesbaden, 1993, Cap. 4.

[76]Hückel, E.: *Ein Gelehrtenleben*, pp. 81–82.

[77]Paul Bernays also attended this lecture. He reworked his notes under the title "*Nature and Mathematical Knowledge*," lectures by David Hilbert, delivered 1919–1920 in Göttingen. Bernays's notes were published in 1992 by Birkhäuser with an introduction in English by David E. Rowe. Cf. Hilbert, D.: *Natur und mathematisches Erkennen*. Birkhäuser, Basel 1992. Hückel's notes from this lecture have been preserved among his papers (SBPK, Hückel papers, box 5, file 2.16). It agrees substantially with Bernays's text.

[78]Hückel, E.: *Ein Gelehrtenleben*, p. 82.

importance of the axiomatic method (axiomatization) in these two fields of knowledge. His point of departure was the fundamental assertion that the basic theory of a discipline can be seen as a system of ideas, concepts or statements about a certain architectonic structure. So some key concepts or theorems of the theory can be regarded as fundamental principles from which the rest, or better put, the structure of the entire system is constructed according to logical principles. Once set up, these fundamental principles can be regarded as an initial point of departure (standpoint), as the "axioms of the individual fields of knowledge." Hilbert names a few such cases from mathematics as well as physics, but also mentions that this selection of principles is only preliminary. In fact, the need arises within the relevant fields of knowledge to base the initial basic principles (axioms) themselves on other more fundamental theorems. Thus theorems that had initially seemed appropriate as axioms are revealed as merely preliminary first principles. The procedure of axiomatization comes about only in the next step when the researcher associates the initial basic theorems to an even lower axiomatic level. Hilbert described the core idea of his axiomatic thinking, as David Rowe aptly remarks, with an architectonic metaphor. He referred to a "relaying of the foundations deeper down."[79] In a later chapter it will become clear that Hilbert's axiomatic thinking had an indirect effect on Hückel's theoretical conceptions when he took up the problem of how to treat the benzene issue quantum theoretically.[80]

During the time that Hückel worked as Hilbert's auxiliary assistant, this influential mathematician offered generally designed lectures on the special and general theories of relativity. As mentioned earlier, Wiechert had questioned Hückel on relativity theory during his doctoral examination. It is hence very probable that Hückel's attitude during the examination contributed significantly to Hilbert's decision to engage him. Hückel's responsibilities were to make himself available as his consultant in the preparations for Hilbert's lectures. In his autobiography Hückel describes his duties and the general ambiance:

> My work with Hilbert was preparing the lectures from discussions with him. These discussions took place in the morning in his office or, in fine weather, in the garden. There he had a large blackboard with a roof on it, and his little dog Peter was our trusty companion. First Hilbert usually started to talk about the political situation, whereby he always came up with original ideas and views. Then the preparations and discussion of the coming lectures were begun. Afterwards I had to rework the lecture material. As a result of my insufficient mathematical skills – which stemmed to a large part from the introduction of the trimester system – I did not always manage, even though Hilbert's assistant of mathematics, Dr. Bernays, sometimes helped me out.[81]

The aim of Hilbert's lecture was to publicize the basic concepts of the special and general theories of relativity among a wider university audience. Erich's brother Walter was among this audience.

---

[79] Cf. Rowe, D. *Introduction*, in: Hilbert, D.: *Natur und mathematisches Erkennen*. Birkhäuser, Basel, 1992, pp. VII–XXIV.

[80] See subsection, Chap. 2.2.2 and 2.2.3

[81] Hückel, E.: *Ein Gelehrtenleben*, p. 82.

It was probably around this time that Erich recommended Ernst Mach's critical history of mechanics: *The Science of Mechanics* to his brother Walter.[82] Mach's focus was on the essentials of the laws of mechanics as well as the limits of their applicability to other fields of knowledge. Walter mentions in this regard in his memoirs: "The critical historical image that Mach draws of mechanics, his acute distinctions between true proofs and apparent proofs first allowed me to understand mechanics properly."[83] Erich, too, probably found in Mach's book an aid to a more profound understanding of the theory of relativity. For, Mach's critical approach toward mechanics was one of the mainstays of Einstein's theory.[84]

Hilbert emphasized in his series of one hour lectures that Einstein's theory of relativity demanded a special intellectual effort, because it set forth how the conceptions of space and time from classical mechanics are invalid. So they had to be replaced by something new that is only graspable by abstraction and thinking by analogy.[85] Thus the Hückel brothers learned from Hilbert's lecture course that more abstract thought structures than the classical conceptions of space and time were necessary, if one wanted to arrive at a deeper insight into the laws of nature. Hilbert's lecture also presented the importance of relativity theory in other sciences beyond physics. A repercussion of Hilbert's message was that the Hückel brothers were sensitized to approaching the foundations of other sciences critically as well. Erich in particular continued to encounter Hilbert's epistemological considerations and ideas.

1.7.2 Assistant to Max Born

When the year as Hilbert's auxiliary assistant came to an end, Hückel was still unable to join Debye in Zurich, because of the same restrictive Swiss laws mentioned above. But another surprise lay in store for him in his home town. The theoretical physicist Max Born, who had returned to Göttingen in 1921, offered him a full-fledged assistantship at his institute starting in the summer term 1922. In his autobiography, Hückel sounded out Born's decision as follows:

> This is as inexplicable to me as Hilbert's decision to engage me as auxiliary assistant. Originally, I was aspiring for experimental physics, not theory. My love of the former may have been dampened by the ordeals I experienced with my doctoral thesis. But perhaps I do have an unconscious aptitude for mathematical thinking. When I was once discussing something with Debye, he told me: "You think far too mathematically." To which I retorted: "Mathematics is beyond me." His answer was: "That's beside the point." W. Pauli also said

[82] This fact is mentioned in Walter's memoirs on p. 212.

[83] Hückel, W.: *Memoiren*, p. 212.

[84] Cf. Holton, G.: *Mach, Einstein, and the Search for Reality*, in: *Thematic Origins of Scientific Thought: Kepler to Einstein*. Harvard University Press, Cambridge, MA, 1988, pp. 237–277.

[85] Cf. Rowe, D.: *Einstein Meets Hilbert: At the Crossroads of Physics and Mathematics*, in: Physics in Perspective 3 (2001), 379–424.

to me once: "One can always have a reasonable discussion with you about all sorts of theoretical things, but you don't come up with anything." I suppose I was too timid to attack problems of unforeseeable scope.[86]

The context was this: after completing his doctorate in July 1921, Wolfgang Pauli started working as Born's assistant in Göttingen that October on Sommerfeld's recommendation. His collaboration with Born was terminated at the end of that winter term by an offer to become Wilhelm Lenz's assistant in Hamburg, one of Arnold Sommerfeld's pupils. Hilbert probably recommended Hückel to Born as Pauli's successor. It is equally likely that Hückel's performance at Born's seminar on "The structure of matter"' during the summer of 1915 had impressed his teacher enough to choose him later as his assistant.

Around this time the quantum theory developed by Bohr and Sommerfeld was applied with relatively good success to explain the band spectra of various polyatomic molecules. The theoretical physicist Kratzer had examined the quantum conditions of the band spectra of diatomic hydrogen halides. He had already shown that in order to explain the finer regularities of the band spectra of the aforementioned heteropolar molecules, one had to take into account the interaction between the rotation of the molecule and the oscillations of its atoms.[87]

In collaboration with Pauli, Born had outlined a procedure by which to calculate this interaction by means of a perturbation calculation for diatomic as well as polyatomic molecules. It was basically an extension of Kratzer's analyses and a continuation of Born's own research, that he had just published with the Hungarian physicist E. Brody.[88] Born wrote to his friend Einstein in Berlin about this on April 31, 1922:

Pauli is unfortunately gone, in Hamburg with Lenz. We recently started a joint paper, a continuation of the one published with Brody on the quantization of anharmonic oscillators. One can apply the approximative procedure developed there to all systems in which the "unperturbed" system is conditionally periodic and the perturbation function can be expanded into a series of powers of one of the parameters. [...] We have also started calculating orthohelium (2 coplanar electrons). [...] Pauli took the paper along to Hamburg and wants to finish it there as I don't have any time for it because of the encycl. article.[89]

Born was working on a comprehensive survey article for Felix Klein's *Encyklopädie der mathematischen Wissenschaften.* It originated from Born's main field of research, the dynamics of lattices in solids.[90] While Born was occupied with

---

[86]Hückel, E.: *Ein Gelehrtenleben*, p. 84.

[87]Kratzer, A.: *Die ultraroten Rotationsspektren der Halogenwasserstoffe*, in: ZP 3 (1920), 289–307; *Eine spektroskopische Bestätigung der Isotopen des Chlors*, in: ZP 3 (1920), 460–465.

[88]Born, M., Brody, E.: *Über die Schwingungen eines mechanischen Systems mit endlicher Amplitude und ihre Quantelung*, in: ZP 6 (1921), 140–152; Born, M. und Pauli jr., W.: *Über die Quantelung gestörter mechanischer Systeme*, in: ZP 10 (1922), 137–158.

[89]Einstein, A., Born, M.: *Briefwechsel 1916–1955*. Nymphenburger Verlagshandlung, München, 1969, p. 102.

[90]Born, M.: *Atomtheorie des festes Zustandes (Dynamik der Kristallgitter)*, in: *Encyklopedie der mathematischen Wissenschaften*, Band V/3, G. B. Teubner, Leipzig, 1923, pp. 527–781.

this survey article, Hückel was supposed to perform the calculations according to the methods carefully prescribed by Born and Pauli and explicitly derive the formulas for each special case. Hückel reported with regard to this work with Born:

> With Born – unlike with Hilbert – I was not involved in the preparations for his lectures. Our discussions usually took place in the afternoon with a cup of tea and some cake in the large office at his private apartment. They mainly concerned the research just mentioned. Two big grand pianos stood in Born's office. My brother Rudi often played chamber music with Born.[91]

During this period Hückel was generally occupied with tedious calculation work. He later recalled: "That was one long string of calculations, which I performed accurately but without enjoying it particularly."[92] Interestingly enough, Born agreed. On August 6, 1922 he wrote his friend Albert Einstein in Berlin about his collaboration with Hückel:

> Nothing much is going on scientifically. I and my assistant Hückel are agonizing over the quantization of polyatomic molecules to calculate the infrared bands (e.g., H_2O). We have the right approximative procedure but the calculations are very complicated. My article for the encyclopedia will probably be finished this month; I am sick and tired of it.[93]

Hückel brought his theoretical paper to successful completion, despite the complicated number-crunching, and published it jointly with Born under the title "*On the Quantum Theory of Polyatomic Molecules*" in the *Physikalischen Zeitschrift*.[94] It is worth mentioning here that during this time Born and Debye were the editors of that physics journal and Hückel was assigned copy-editing tasks. This may explain why the paper could be published so speedily and smoothly.

As Helge Kragh has aptly described, Hückel obtained from Born his first demanding theoretical assignment along with his first dabblings with quantum theory in a new peripheral field situated between physics and chemistry. A few years later it acquired the neologism "chemical physics."[95] As Debye was later glancing through the correction proofs, his remark to Hückel was: "This looks like Born."[96] The quantum-theoretical considerations on polyatomic molecules by Born and Hückel certainly do mirror Born's mentality and stylistic orientation as a scientist. Following the mathematical tradition at Göttingen, Born tended rather to rigorous mathematical proofs of familiar physical concepts and ideas than to the discovery of new ideas.[97]

---

[91] Hückel, E.: *Ein Gelehrtenleben*, p. 85.

[92] Ibid.

[93] Einstein, A., Born, M.: *Briefwechsel 1916–1955*, p. 106.

[94] Born, M., Hückel, E.: *Zur Quantentheorie mehratomiger Molekeln*, in: PZ 24 (1923), 1–12.

[95] Kragh, H.: *Before Quantum Chemistry: Erich Hückel and the Physics-Chemistry Interface*, in: CENTAURUS 43 (2001), 1–16.

[96] Hückel, E.: *Ein Gelehrtenleben*, p. 85.

[97] Cf. Staley, R.: *Max Born and the German Physics Community: The Education of a Physicist*. PhD thesis, Cambridge University, UK, 1992.

Molecular spectra were an area in which, in some respects, the Bohr–Sommerfeld quantum theory hit upon its limits. For example, Pauli's dissertation attempted to apply this theory to the hydrogen molecule ion (H_2^+).[98] The result was disappointing because the theory did not arrive at the correct stationary states. What is more, the results of Pauli's dissertation raised doubts about the validity of the Bohr–Sommerfeld model.[99] Bohr's and Sommerfeld's quantum theory was evidently unsatisfactory in some respects.

The lectures that the Danish theoretician Niels Bohr delivered in Göttingen between June 12 and 22, 1922 at Hilbert's invitation to the seminar for mathematics and physics turned into a momentous event for science. With subtlety and lucidity Bohr surveyed the current state of the quantum theory of atoms and managed to introduce the first somewhat satisfactory explanation for the periodic system of the chemical elements. The members of his audience included such prominent figures of the Göttingen scientific community as the mathematicians Richard Courant, Hilbert and Carl Runge, as well as the physicists James Franck, Born and Robert Pohl. Important specialists from elsewhere in Germany also traveled to Göttingen for the occasion. Sommerfeld appeared from Munich with his doctoral student Werner Heisenberg; Alfred Landé, Erwin Madelung and Walter Gerlach came from Frankfurt; Lenz and his assistant Pauli arrived from Hamburg and Paul Ehrenfest crossed the Dutch border from Leyden. Bohr's lectures before roughly a hundred listeners have gone down in history as "The Bohr Festival."[100] Born's assistant at that time, Friedrich Hund, who was working on his dissertation, remembered the general mood among the younger members of the audience:

> Bohr delivered lectures on Mondays, Tuesdays and Wednesdays, at the time of the regular seminar (usually lasting considerably longer) over the course of three weeks, on the quantum theory of atoms and the periodic system of the elements. Bohr did not speak clearly, and we younger ones were not allowed to sit on the front benches among the important guests; so we listened intently with craning ears, fighting our suppertime appetites. We had certainly read something of Sommerfeld's *Atombau und Spektrallinien*; 1920 Debye also delivered a lecture on quantum theory (in a quite unheated auditorium); but what Bohr lectured on sounded quite different and we sensed that it was something quite essential. The brilliance surrounding that event cannot be reflected today; for us it was as magnificent as the Handel Music Festival in Göttingen of that period.[101]

Hückel was one of these younger auditors, together with Franck's assistant Rudolph Minkowski, who both had the difficult task of taking notes of Bohr's

---

[98] Pauli, W.: *Über das Modell des Wasserstoffmolekülions*, in: Annalen der Physik (4) 68 (1922), 177–240.

[99] For more details, see: Jensen, C.: *Two One-Electron Anomalies in the Old Quantum Theory*, in: Historical Studies in the Physical Sciences 15 (1984), 81–106.

[100] *The Bohr Festival in Göttingen*, in: Mehra, J., Rechenberg, H.: *The Historical Development of Quantum Theory*, vol. 1, Part 1. Springer-Verlag, New York Heidelberg Berlin, 1982, Chapter III, pp. 259–358.

[101] Hund, F.: *Göttingen, Kopenhagen, Leipzig im Rückblick*, in: Bopp, F.: *Werner Heisenberg und die Physik unserer Zeit*. Vieweg, Braunschweig, 1961, pp. 1 f.

lectures and reworking them into a finished text.[102] Bohr's lectures at Göttingen certainly had a decisive effect on the scientific careers of the younger generation of physicists, indeed on the entire physics community at Göttingen. They inaugurated the major developments in quantum theory that were conceived in that university town and beyond within Germany.

Pauli, for instance, left for Copenhagen soon afterwards at Bohr's invitation to work with him for a year. He returned to Hamburg in the fall of 1923 and continued to work on the stimulating ideas he had gained in Copenhagen in his efforts to provide a theoretical interpretation of the periodic table of the chemical elements. Toward the end of 1924 he discovered the "exclusion principle" that is now also known as the Pauli principle. Two years later Pauli succeeded in calculating the stationary states of the hydrogen atom and its spectrum using the new matrix mechanics developed by Born, Pascual Jordan and Heisenberg. This achievement bolstered confidence in the new mechanics.

Born, for his part, began in January 1923 to study the intricate methods of celestial mechanics Henri Poincaré had employed. Together with his collaboraters at Göttingen, who included Heisenberg, Hund, Jordan and Lothar Nordheim, his aim was to apply these methods to problems in the physics of the atom. In a letter dated April 7, 1923 Born confided to his friend Einstein:

> Despite every effort I cannot get near the grand quantum puzzle. We studied perturbation theory (per Poincaré), to establish whether upon careful calculation from the Bohrian models one arrives at the observed term values. But this is *quite* certainly *not* the case, as has been demonstrated with helium, where we found all possible multiply periodic orbits (to sufficient approximation). I had Heisenberg here during the winter (since Sommerfeld was in America). This man is at least as talented as Pauli, but nicer and more pleasant personally. He also plays the piano very well. Besides collaborating on the helium paper, we also examined a few principal questions of Bohr's atomic theory, particularly on the phase relations in models of the atom.[103]

The results of these investigations were subsequently published in articles by Born and Heisenberg on the excited states of the helium atom[104] and by Nordheim on the quantum theory of the hydrogen molecule ion.[105] Both publications clearly established the failure of the prevailing theories of the atom. It finally prompted Born's group of researchers to design a new mechanics of quanta.

How was the young Hückel affected by Bohr's lectures and the subsequent stimulating discussions among Born's close colleagues? His negative number-crunching

---

[102] AHQP, Interview with R. Minkowski, April 1962, p. 1. For more details, see: Bohr, N.: *Collected Works*, Band 4. North-Holland Physics Publishing, Amsterdam – New York – Oxford, 1977, pp. 341–419.

[103] Einstein, A., Born, M.: *Briefwechsel 1916–1955*. Nymphenburger Verlagshandlung, München, 1969, p. 109.

[104] Born, M., Heisenberg, W.: *Die Elektronenbahnen im angeregten Heliumatom*, in: ZP 16 (1923), 229–243.

[105] Nordheim, L.: *Zur Quantentheorie des Wasserstoffmolekülions*, in: ZP 19 (1923), 69–93.

experience had apparently discouraged him from continuing to work on atomic theory. It is also notable that Hückel did not pursue any independent research related to quantum theory in this period. He only tackled a problem Born had assigned him. Having taken his degree in experimental physics there was still much to be learned in the field of atomic and quantum theory before he could participate actively in its further development.

Meanwhile, the employment restrictions in Switzerland had been lifted and Hückel received Debye's invitation to work as his assistant in the fall of 1922. Thus Hückel was finally able "to realize his original plan and join Debye."[106] He later justified his decision to accept Debye's offer as follows:

> I immediately accepted, on one hand, because I had spent my entire youth and period of studies at Göttingen and also wanted to become acquainted with something else and, on the other, because the work with Born did not particularly appeal to me. I couldn't know that the new quantum mechanics was in the offing for Born's group of collaborators. Nor would I have dared to interfere actively. My mathematical skills and knowledge probably would not have sufficed. [...] I could not have delivered what Born had to demand, or what would have been necessary, in order to participate in the development of quantum mechanics.[107]

Hückel's description of his missed opportunity to collaborate on the development of the new quantum mechanics in Göttingen is very modestly put. But he could hardly have foreseen the development of the new quantum mechanics now bearing Born's and Heisenberg's names. As mentioned earlier, Born, Heisenberg (who was Hückel's successor for a semester), Nordheim, Hund (Heisenberg's successor), and the other colleagues at Born's institute started to discuss the failure of the "old quantum theory" in the spring of 1923. The decisive breakthroughs came in the subsequent three years.

Without a doubt Hückel needed a fresh start after his latest experience with Born in Göttingen. This opportunity came at Debye's physical institute at the ETH in Zurich. His collaboration there started a new phase in his scientific career.

1.8 Assistantship Under Peter Debye in Zurich and Habilitation Thesis on the Theory of Strong Electrolytes

When Hückel arrived at the physical institute of the *Eidgenössische Technische Hochschule* (ETH) in the fall of 1922, Debye immediately set him to work on a new project in the field of physical chemistry: it concerned the theory of strong electrolytes. A theory of electrolytic dissociation had been enunciated by Svante Arrhenius (1859–1927) and Wilhelm Ostwald (1853–1932) – the founding fathers of physical chemistry together with Jacobus Henricus van't Hoff (1852–1911). It states that strong electrolytes, such as table salt (NaCl) dissolved in water dissociates

---

[106] Hückel, E.: *Ein Gelehrtenleben*, p. 85.

[107] *Chemiker im Gespräch: Erich Hückel*, in: Chemie in unser Zeit 4 (1970), 180–187, p. 181.

into positively charged (Na^+) ions and negatively charged (Cl^-) ions. This dissociation is not complete because a portion of the salt dissolves into electrically neutral molecules (NaCl). This assumption prompted various predictions about the behavior of aqueous solutions of strong electrolytes, for instance, the electric conductance, lowering of the freezing point and related phenomena. The degree of dissociation of an electrolyte and its total concentration at a given temperature could be used to figure out the electric conductance and vice versa. Around the turn of the century, however, experiments revealed various contradictory indications and inconsistencies with the theory. A better explanation of the behavior of strong electrolytes was needed.[108]

Some scientists chose to simply ignore the anomalies and employed formulas derived from empirical observation. But the physics behind such empirical formulas and the corresponding adjustments remained a mystery until the following two new and unexpected insights were made. The Danish chemist Niels Bjerrum (1879–1958) first explicitly mentioned that a complete dissociation of strong electrolytes would be possible. In 1909 he suggested that the conductance and the freezing-point depressions observed by other researchers had to be explained by interactive forces between the ions. Bjerrum's suggestion permitted the British physicist Samuel Milner at the University of Sheffield to develop a mathematical model for the interionic forces between dissolved ions and solvent. Between 1918 and 1921 the Indian chemist Chandra Ghosh published papers in the *Journal of the Chemical Society*. He assumed that even in solution the rigid crystal lattice of a strong electrolyte remained intact. This assumption simplified the calculations for the interionic forces between solvent and solute and led to what is known as Ghosh's equation for the conductance of binary electrolytes.[109]

Inspired by a talk that the Swiss physicist Edmond Bauer had given before the Zurich Physical Society about Ghosh's work, Debye laid the foundation of the later theory of strong electrolytes in the winter of 1921, already before Hückel's arrival:

> This man Bauer, he comes now and gives a talk. And he starts his talk by saying, "Now we have long had old shoes" – I still remeber that – "old shoes in physical chemistry, to walk on. But now we have a pair of good new ones. And there is an Indian who has delivered them, an Indian by the name of Ghosh." His picture was that you take a sodium chloride solution, then the sodium and the chloride in the solution are in the same position with respect to each other as in a crystal. With increasing dilution, the distances get bigger and bigger. And from that he derived a third root law, you see third root of the volume, you

---

108This issue is too involved to permit a summary of the details within the limits of a few pages. Cf. Servos, J. W.: *Physical Chemistry from Ostwald to Pauling, the Making of a Science in America*. Princeton University Press, Princeton, NJ, 1990. Chapter 3, *From Processes to Structures: The Anomaly of Strong Electrolytes and the Problem of the Chemical Bond*, pp. 120–150; Falkenhagen, H.: *Elektrolyte*. Hirzel, Leipzig, 1932.

109For further historical details and pertinent bibliographies, see Wolfenden, J. H.: *The Anomaly of Strong Electrolytes*, in: Ambix 19 (1972), 175–196; Snelders, H. A. M.: *The Historical Background of Debye and Hückel's Theory of Strong Electrolytes*, in: Proceeding of the Royal Netherlands Academy of Arts and Sciences B 89 (1986), 79–94.

> know? So that was what the new shoes were. And Nernst believed in that. And so in this place I said, that cannot be possible, because these things must have Brownian motion. And I got in trouble with them, and we were a little bit heated up over that. (They) said, "Well, if you really mean that, you will have to do something about it!" So I went home and did something! That was all, you see. But then I did not know about the literature. So I said to Hückel who was my assistant. "Now you, boy, you study the literature and tell me about it so then we`ll see what we are missing." That`s how the whole thing came about.[110]

So in Zurich Hückel once again started working on an assigned task: He was supposed to take up where Debye had left off and develop the initial ideas, calculate them out, and discuss the results with Debye.[111] The inquisitive assistant succeeded in mastering the new theoretical problem quite quickly. The first research results by Debye and Hückel on the new theory of strong electrolytes was submitted to the *Physikalischen Zeitschrift* at the end of February 1923 and was published on May 1, 1923. A few months later, on August 1, 1923 the second communication appeared in the same journal.[112] It is unnecessary to retrace the mathematical arguments underlying Debye's and Hückel's ideas.[113] I will just emphasize a few general considerations. Debye and Hückel analysed and simplified Milner's model and found new functions for the ion distribution on the basis of statistical thermodynamics. They found a simpler square root law in replacement of Chandra Ghosh's "third root law." As a result, the Debye–Hückel theory agreed on the whole with the experimental measurements of the freezing-point depressions and the limiting law of the electric conductance in dilute solutions. The new hypothesis of complete dissociation of strong electrolytes was also fully confirmed by the new theory. The Norwegian physical chemist Lars Onsager (1903–1976) then extended the Debye–Hückel model to nonaqueous solvents. Onsager, who took a special trip from Oslo to Zurich in order to discuss the problem with Debye, improved the theory of the conductance of electrolytic solutions. Hückel remembered Onsager's visit to the physical institute in Zurich:

> One morning in the spring of 1925 someone knocked on Debye's office door in Zurich while we were in the midst of a discussion. Prompted by the "Come in!", a very young, tall, blond man walked in through the doorway. To Debye's asking: "What do you want?" came the blunt response: "Your theory is wrong." This referred to the theory of the electric conductance of strong electrolytic solutions that Debye and I had published. There was, in fact, an incorrect slip-up that falsified the result somewhat.[114]

In his considerations Onsager took into account that the central ion itself was subject to Brownian motion. Debye and Hückel had merely assumed a linear motion

---

[110]Original English. AHQP, Interview with P. Debye, May 3, 1962, p. 18.

[111]*Chemiker im Gespräch: Erich Hückel*, p. 181.

[112]Debye, P., Hückel, E.: *Zur Theorie der Elektrolyte. I. Gefrierpunktserniedrigung und verwandte Erscheinungen*, in: PZ 24 (1923), 185–206; *Zur Theorie der Elektrolyte. II. Das Grenzgesetz für die elektrische Leitfähigkeit*, in: PZ 24 (1923), 305–325.

[113]Cf. Falkenhagen, H.: *Elekrolyte*. Hirzel, Leipzig, 1932.

[114]Hückel, E.: *Ein Gelehrtenleben*, pp. 110–111.

in their approximation. Thus the calculations and considerations Onsager published between 1926 and 1927 yielded considerably better agreement with experiment.[115]

The Debye–Hückel theory of electrolytic solutions clearly mirrors Debye's characteristic thinking and scientific approach. Debye could immediate narrow in on the crux of a problem and then formulate it appropriately in order to describe the related processes by the method of approximation. The final result of his calculations was then compared against the pertinent experimental data. Hückel fittingly described Debye's thinking and working approach in his memoirs:

> Debye saw – as it were, intuitively – the physical essentials and then found a simple way to support this insight and to solve the problem at hand mathematically in a relatively simple way. [...] The aforesaid already reflects the basics of Debye's approach. Debye tackled his research – in my view – like an artist who labors out of sheer pleasure in the topic and the exercise, letting himself often be guided by intuition, which he retroactively underpins rationally as simply and clearly as possible and eliminating whatever is superfluous. [...] Debye's thinking was entirely different from A. Einstein's. Debye did not have such philosophical prejudices, either, as are found in Einstein's downright dogmatic obstinacy about a statistical interpretation of quantum theory: "God throws no dice" [...], at least I have never noticed the slightest interest by Debye in philosophical questions.[116]

So Debye had a keen eye for the simple essentials. Debye's own self-image also incorporates an emphasis on intuitive insight into the holistic structures of a problem. It was not mathematics but thinking in pictures, as he put it in retrospect, that took first place in his considerations:

> I have always felt that you cannot do without a picture. If you talk about the hydrogen atom, you have to start with the potential energy between an electron and a nucleus. Then you express it in the form of a Hamiltonian, and so forth. But you have not avoided the picture. The question is whether the model is the main thing, or whether the mathematical handling of this picture is the main thing. Nowadays, there is a lot of emphasis on the mathematical manipulation. I think that is all right, but I cannot do without a picture. (...) Mathematicians have been saying that something which is simple in mathematics has to have an application in physics. Of course, this is an extreme position. But this is a personal matter: whether one starts with a picture and then tries to make a formulation which represents all the experiments, or if one thinks of a mathematical formulation which looks nice to him, and then looks to see if he can get a physical interpretation. There must then be a confrontation of these two parts.[117]

The picture that Debye started out with in his theory of electrolytes and from which he elaborated the physics mathematically in collaboration with Hückel was a spherical ion surrounded by an "ionic atmosphere" of opposite charge and moving linearly under the influence of an electric field.

In retrospect, Hückel's "apprenticeship" under Debye can be regarded as the period of methodological preparation for his own pioneering accomplishments in

---

[115] Onsager, L.: *Zur Theorie der Elektrolyte. I*, in: PZ 27 (1926), 388–392; *Zur Theorie der Elektrolyte. II*, in: PZ 28 (1927), 277–298.

[116] Hückel, E.: *Erinnerungen an Peter Debye und an meine Lehrjahre*, in: Physikalische Blätter 28 (1972), 53–57, p. 54 f.

[117] Original English. Debye, P.: *An Interview*, in: Science 145 (1964), 554–559, pp. 554–555.

the forming border discipline of quantum chemistry.[118] Debye later explained about his own apprenticeship with Sommerfeld that he had not learned physics but "the mathematical formulation of the physics."[119] Hückel, for his part, learned three things in particular from Debye: First, how to identify important and controversial problems still awaiting satisfactory solution. Second, how to draft a satisfactory theory on the basis of bold assumptions and approximative methods leading to plausible explanations of specific experimental data.[120] Third, how to chisel at his skills in mathematics and physics in order to have the tools to carve out suitable models to explain diverse physical and chemical phenomena. The Debye–Hückel theory of strong electrolytes was a fine example of a classical theory constructed on the basis of familiar laws and without any ad hoc assumptions. It was grounded in electrodynamics, thermodynamics and statistical physics but not on obsolete forms of quantum theory or models of the atom.[121]

Chemists and physical chemists were initially far from enthusiastic about Debye's and Hückel's theoretical arguments. The majority of them were too comfortable with the theoretical framework underlying Arrhenius's theory of partial dissociation. In 1924 Hückel published a survey article on the subject of electrolytes in the journal *Ergebnissen der exakten Naturwissenschaften*.[122] It triggered a confrontation with the influential physical chemist Walther Nernst (1864–1941), who had been upholding a controversial point of view for a long time. Hückel had vivid recollections of his hefty dispute with Nernst.

> Soon afterwards I met Nernst at a conference. Nernst and I becamed engaged in a conversation during which he proceeded to give me a sound dressing down over a bottle of champagne. He said I had gone too far with my representations, had written much nonsense and ought to revoke. He would not believe that strong electrolytes could be regarded as completely dissociated but that an undissociated portion always had to be taken into account, especially when treating the equilibrium of a saline solution. I refused to retract my claims. A continuation of our conversation did not take place, because Nernst did not appear at the agreed time and place. At the foregoing confrontation Nernst had threatened: "A storm is soon going to break over you!" I replied coolly: "Herr Geheimrat, storms do pass."[123]

The sensitive and timid young man evidently had no qualms about confronting influential authorities when it came to defending his own scientific views. Hückel did not have to weather the threatened "storm." Nernst incorporated the Debye–Hückel theory soon afterwards in the new edition of his book *Theoretische Chemie* from 1926.[124] A few years later Nernst and his collaborator Wilhelm

---

[118]Cf. Hartmann, H., Longuet-Higgins, H. C.: *Erich Hückel*, p. 156.

[119]Original English. AHQP, Interview with P. Debye, May 4, 1962, p. 5.

[120]Cf. Berson, J. A.: *Erich Hückel – Pionier der Organischen Quantenchemie: Leben, Wirken und späte Anerkennung*, in: Angewandte Chemie 108 (1996), 2922–2937, p. 2924.

[121]Cf. Kragh, H.: *Bevor Quantum Chemistry*, p. 11 and p. 14.

[122]Hückel, E.: *Zur Theorie der Elektrolyte*, in: Ergebnisse der exakten Naturwissenschaften 3 (1924), 199–276.

[123]Hückel, E.: *Ein Gelehrtenleben*, p. 93.

[124]Kragh, H.: *Before Quantum Chemistry*, p. 11.

Orthman even conducted experimental research that explicitly took their new theory into account.[125]

Hückel's responsibilities as assistant in Zurich were not confined to pure research. He helped the laboratory assistant set up the demonstration experiments for Debye's physics lectures and acted as a consultant and aid to the students during the practical laboratory sessions. When Debye was invited to deliver a series of lectures in the U.S. over a period of a few weeks, he entrusted Hückel with the main experimental lecture. The relations between Debye and his assistant were uncomplicated and, indeed, quite unconventional. Hückel later described his teacher's personal style: "My personal relationship with Debye was always completely untroubled. Although I was 12 years younger than him and an utter novice when I joined him in Zurich, he always treated me as his equal."[126] The young Hückel was fascinated by Debye not only as a genious researcher but also as a person able to enjoy life:

> He liked to eat well: Once, for instance, when a problem could not be solved during the discussion after a physics colloquium, he said: "A good, hearty supper is what we need here, one can always think of something then."[127]

The young Hückel also enjoyed life in Zurich. He was a frequent guest at the open-air swimming pool on the shore of Zurich Lake. But physics was never very from his mind, as the following quote illustrates: "sometimes the water temperature rose to 27°C." Excursions, steamboat trips and dance parties were organized at the institute. After such a break, he could resume his work on his habilitation thesis with redoubled energy.

In December 1924 he completed his thesis and it appeared in the January issue of the *Physikalischen Zeitschrift* in 1925. The subject was the theory of concentrated aqueous solutions of strong electrolytes.[128] This thesis extended the theory of strong electrolytes beyond very dilute solutions to high concentrations. The conferral of his habilitation degree also had advantages for his personal life. It comprehended a promotion to first assistant with a raise in salary from 300 to 500 francs. He could thus afford to marry the love of his life, Annemarie. Her father Richard Zsigmondy (1865–1929) was the Nobel laureate in chemistry for the year 1925. He had set this academic qualification as a condition for consenting to his daughter's wedding.

Hückel had met Annemarie Zsigmondy during his student days in Göttingen. Erich and his brother Walter were close friends of the chemistry student Paul Schumm, who brought the future couple together. Hückel remembered this important event of his life in his autobiography:

---

[125]Nernst, W., Orthman, W.: *Die Verdünnungswärme von Salzen bei sehr kleinen Kozentrationen*, in: ZPC 135 (1928), 199–208.

[126]Hückel, E.: *Ein Gelehrtenleben*, p. 56.

[127]Ibid., p. 57.

[128]Hückel, E.: *Zur Theorie konzentrierter wässeriger Lösungen starker Elektrolyte*, in: PZ 26 (1925), 93–147.

> Professor Kötz was extraordinary professor of inorganic chemistry at Göttingen and Schumm was one of his students. The wife of this professor used to invite a group of his students on Saturdays for afternoon coffee. My future wife Annemarie was acquainted with Mrs. Kötz through her mother, wife of the professor of colloidal chemistry Zsigmondy. That is how Annemarie joined the circle at Mrs. Kötz's, and she met Schumm there. Schumm played the piano well and Annemarie the violin. His music-making gained him access to the Zsigmondy household and my acquaintance with him mine. Thus I got to know Annemarie and Professor Zsigmondy and our conversations and discussions did not remain at the purely personal level, extending also to scientific topics of mutual interest.[129]

The civil ceremony of Ernst Hückel and Annemarie Zsigmondy took place on August 8, 1925 and their church wedding with subsequent private celebrations in the apartment of the bride's family ten days later, on August 18. The new couple left for their honeymoon the following day. They traveled to Würzburg, Munich and Pertisan on Lake Achen. Continuing on to Terlago near Trento, where Annemarie's father owned a large old manor house, they spent the remainder of their honeymoon in natural surroundings that were then still pristine.

The young Hückels returned from Terlago for the beginning of the winter term 1925/26. Erich had rented an cramped three-room apartment on Rotbuchstrasse on the outskirts of the city and furnished it with their few belongings. This is how he described it:

> Our furnishings were frugal. The wealth of our parents had completely vanished in the rampant inflation in Germany. So Anne's mother had bought the bare essentials from auctions. Even so, our apartment turned out to be quite cozy.[130]

In Zurich Annemarie embarked on her new social life and Erich on a new phase of his researches.

1.9 From the Theory of Strong Electrolytes to the Quantum Theory of Double Bonding

1.9.1 New Ambitions and Separation from Peter Debye

Even before the wedding, Hückel and his father-in-law, the colloidal chemist Zsigmondy, had been exchanging thoughts on scientific issues of interest to both of them: problems in colloidal chemistry. Hückel tells us, very modestly: "He told me many times that he esteemed my scientific expertise very highly. – A little too highly, I think."[131]

These discussions produced a short coauthored article on the growth of gold particles in colloidal auriferous solutions that appeared in June 1925 in the *Zeitschrift*

---

[129] Hückel, E.: *Ein Gelehrtenleben*, p. 87. Hückel's mother-in-law, Laura Luise was one of the daughters of the professor of anatomy at Jena, Wilhelm Müller.

[130] Ibid., p. 110.

[131] Ibid., p. 105.

für physikalische Chemie.[132] Thus Hückel's interests ventured beyond the theory of electrolytes into the territory of colloidal chemistry.

In the winter term 1925/26 Hückel embarked on a new project. He began with intense concentration to write a book on adsorption and capillary condensation of gases and vapors on solid surfaces and porous bodies. It was published a few years later in a series called "Individual portraits of colloidal research" initiated by his father-in-law and edited by Herbert Freundlich.[133] It was one of Hückel's priorities to treat the thermodynamics in this area with special thoroughness because the related issues had hitherto received rather superficial attention.[134] Upon receiving Hückel's book, Debye wrote him:

> Thank you very much for kindly sending me your book. I glanced through many different parts of it and was always entirely pleased with its comprehensiveness and clarity. I occasionally felt a pang of pity for you, when I saw how much work you had to invest in the book. Now that I have seen how good and nice a thing it has turned out to be, I am content again.[135]

Hückel intended to write a second volume on adsorption from solutions. With Debye's recommendation he applied for a stipend from the International Education Board to visit the biochemist Frederick George Donnan (1870–1956) at University College in London for this purpose. Donnan was one of the founders of physical chemistry in England. But Hückel fell sick in the summer of 1926 and had to delay these plans. He suffered from a serious bout of appendicitis with attendant irritability of his entire digestive tract. Hückel was forced to take sick leave from his position in Zurich to treat his generally weakened state of health, first in Göttingen, then in a private clinic in Frankfurt. His research was at a standstill for months. This was not without repercussions on his career.

Debye received an appointment as full professor of experimental physics at the University of Leipzig in 1927, to fill Otto Wiener's chair. Debye was to revive that university's tradition going back to the great physical chemist Ostwald. He established Leipzig as an international center of research in physical chemistry and theoretical physics. Debye's move away from Zurich meant a parting of teacher and pupil. Hückel decided to leave for London to work with Donnan. He wrote to Debye on November 12, 1927 from Zurich: "I have decided to go to London after all. [...] I think that I will learn interesting new things there."[136] It is not known whether Hückel had been offered the chance of following his mentor to Leipzig. In

[132] Zsigmondy, R., Hückel, E.: *Über Reduktionsgeschwindigkeit und das Wachstum kleiner Goldteilchen bei der Herstellung kolloidaler Goldlösungen*, in: ZPC 116 (1925), 291–303.

[133] Hückel, E.: *Adsorption und Kapillarkondensation*, in: Bd. 7 der Kolloidforschung in Einzeldarstellungen. Akademische Verlagsgesellschaft, Leipzig, 1928.

[134] Hückel, E.: *Ein Gelehrtenleben*, p. 88.

[135] Debye to Hückel, Zürich, 27. Januar 1928, MPG-Archiv, III. Abt., Rep. 19, NL Debye.

[136] Hückel to Debye, Zürich, 12. November 1927, MPG-Archiv, III. Abt., Rep. 19, NL Debye.

any case, Debye took his assistant Heinrich Sack with him from Zurich.[137] But he continued to stay in contact with Hückel over the next 40 years.[138] Hückel offers the following reason for this separation in his reminiscences of his teacher:

> Thus Debye and I parted ways, which had, incidentally, been planned, because Debye thought it not good for my further scientific development if I stayed forever under his wing.[139]

Debye would doubtlessly have very much liked to keep Hückel as his assistant, appreciating him both as a person and a scientist. But we gather from Hückel's narrative how Debye, very kindly and "gently," broke the following news before his departure.

> When everything was settled for his imminent transfer to Leipzig, he said to me once: "For a theoretical professorship I do have to take someone – like Heisenberg, for instance." He apparently did not what to say to me directly "You unfortunately don't come into consideration." It would not have been necessary, either. I knew perfectly well that this would have never been possible for me. My theoretical knowledge and skills were truly much too limited and meagre for the thought to even enter my head.[140]

This "theoretical professorship" referred to the chair for theoretical physics that had become vacant upon the death of Theodor Des Coudres in 1927. As a newly appointed *Ordinarius* for experimental physics and director of the physics institute, Debye immediately had a strong influence on the filling of this vacant position. Des Coudres, who had occupied it since 1902, incarnated the physicist of the old style, feeling at home in classical theoretical physics as well as experimental physics. But things had changed since the dawn of the 20th century, as Wiener wrote Sommerfeld shortly before his death.[141] It was no longer "deemed of critical importance that the person in question also be an experimenter, like Des Coudres. We would rather prefer an outstanding theoretician."[142] The leading names on the list of nominees for Leipzig were Heisenberg and Pauli. Both these top candidates were outstanding

---

[137] One element of Debye's negotiations with the Ministry of Culture in Dresden had been that he be able to take Sack with him to Leipzig. In addition, the Ministry supported a grant application to the Notgemeinschaft der deutschen Wissenschaft for his second assistant H. Falkenhagen. Debye an Ministerialrat von Seydewitz, 19 September 1927; Debye an Notgemeinschaft, 20 Juli 1928, MPG-Archiv, III. Abt., Rep. 19, NL Debye.

[138] Cf. Fleck, G.: *Peter Debye (1884–1966)*, in: *Nobel Laureates in Chemistry 1901–1992*, edited by Laylin James. American Chemical Society and the Chemical Heritage Foundation, 1993.

[139] Hückel, E.: *Erinnerungen an Peter Debye*, p. 55.

[140] Hückel, E.: *Ein Gehlertenleben*, p. 122.

[141] Wiener died a short time after Des Coudres.

[142] Eckert, M.: *Die Atomphysiker. Eine Geschichte der theoretischen Physik am Beispiel der Sommerfeldschule*. Vieweg, Braunschweig/Wiesbaden, 1993, p. 99.

theoreticians who played a pivotal role in the emergence and development of quantum mechanics. The minutes of the negotiations reveal that Debye sought to obtain Heisenberg, who ultimately received the call as well.[143]

As the above quote demonstrates, Hückel was unpresumptuously aware of his own theoretical limitations: He had in the interim lost touch with the dramatically changing developments in quantum theory. Since his departure from Born in Göttingen, his research on the theory of electrolytes, his book on adsorption and particularly his serious illness had shut him out of the new field of theoretical physics, quantum and wave mechanics. Meanwhile a new elite of theoreticians had formed to which Hückel no longer belonged. Indeed, by 1926 Hückel did not even seem to be particularly interested in the new developments of quantum theory. Erwin Schrödinger, who was professor of theoretical physics at the University of Zurich throughout Hückel's assistantship at the Polytechnic, once gave a talk at the colloquium at Debye's invitation towards the end of 1925. He discussed Louis de Broglie's theory of material waves and his own new insights about it.[144] There is no documentation on whether Hückel had figured in any way at Schrödinger's colloquium or even taken part in it. At that time he was fully occupied with his book on adsorption. But he would have had the best possible preconditions to learning about the latest accomplishments in the field of quantum theory from Schrödinger personally. Hückel knew him and his wife very well and, together with Annemarie, had "been their guests many times."[145]

But one thing is certain. During his time in Zurich Hückel was fully aware of the new theory, if only from the difficulties that his colleagues and physics students encountered in trying to understand the physical significance of Schrödinger's mysterious "psi function." Felix Bloch, who was studying physics in Zurich at the time and a few years later produced fundamental contributions on the quantum mechanical electron theory of metals,[146] recalls about the year 1926 in his memoirs in this regard:

---

[143]Cf. Ibid., p. 101. Following old German custom, a candidate was not engaged but "called." He did not even have to have applied for the position. The transmission of knowledge was considered more than a profession, it was a calling. For details on Heisenberg's call to Leipzig, see Cassidy, C. D.: *Werner Heisenberg: Leben und Werk*. Spektrum Akademischer Verlag, Heidelberg Berlin Oxford, 1995, Chap. 12.

[144]Debye, P.: *An Interview*. Debye described his conversations with Schrödinger: "I was at the Technical University, which is a federal institute, and we had a colloquium together. We were talking about de Broglie`s theory and agreed that we didn`t understand it, and that we should really think about his formulations and what they mean. So I asked Schroedinger to give us a colloquium. (...) It was in the same year that he published his paper, because there were only a few months between his talk and his publication" Original English, p. 145. Schrödinger's papers (four communications) were published in the *Annalen der Physik* under the title "Quantization as an Eigenvalue Problem" (*Quantisierung als Eigenwertproblem*) between January and June 1926.

[145]Hückel, E.: *Ein Gelehrtenleben*, p. 126.

[146]Bloch's contribution to the quantum mechanical electron theory of metals and Hückel's other application of this theory for a quantum mechanical interpretation of the electron configuration of benzene and other aromatic compounds, see Chapter 2.2.2.

> Of course, there was afterwards a lot of talk among the physicists of Zurich, including even the students, about that mysterious "psi" of Schrödinger. In the summer of 1926, a fine little conference was held there and at the end everyone joined a boat trip to dinner in a restaurant on the lake. As a young *Privatdozent*, Erich Hückel worked at the time on what is now well known as the Debye-Hückel theory of strong electrolytes, and on the occasion he incited and helped us to compose some verses, which did not show too much respect for the great professors.[147]

The poem by Hückel that Block referred to reads:

Gar Manches rechnet Erwin schon	What a lot Erwin can calculate
Mit seiner Wellenfunktion.	With his wave function.
Nur wissen möcht man gerne wohl	One would just like to know
Was man sich dabei vorstell'n soll.[148]	What ever one is to make of it.

This little verse reflects humorous skepticism rather than any particular enthusiasm towards Schrödinger's new wave mechanics. Hückel's genuine interest in it formed a few years later in a different context, as the following chapter describes.

1.9.2 Fellowship from the International Education Board in London with Frederick Donnan: New Research Orientation

When Hückel finally decided to leave for London, Debye applied the pressure to accelerate the fellowship award they had already solicited from the International Education Board. In the meantime Hückel embarked on a new research project. In a letter to Debye from January 30, 1928 he wrote: "After not quite knowing what specifically to choose at the beginning of the term, I have been working for some time on a theory of the mobility of the hydrogen and hydroxyl ion, which is now essentially finished."[149] Hückel's health condition was quite good right then,[150] allowing him to complete the mentioned paper in the following months. It was published soon afterwards in *Zeitschrift für Elektrochemie und Angewandte Physikalische Chemie*.[151]

During the semester break Hückel went to Göttingen, where he received a telegram from Debye announcing the approval of the grant he sought. He received a stipend for the period of one year as of April 1, 1928 in order to conduct research

---

147 Original English. Bloch, F.: *Reminiscences of Heisenberg and the Early Days of Quantum Mechanics*, in: Physics Today 29 (12) (1976), 23–27, p. 24.

148 Ibid.

149 Hückel to Debye, Zürich, 30. January 1928, MPG-Archiv, III. Abt., Rep. 19, NL Debye.

150 Debye to Hückel, Zürich, 8 März 1928, MPG-Archiv, III. Abt., Rep. 19, NL Debye.

151 Hückel, E.: *Theorie der Beweglichkeit des Wasserstoff- und Hydroxylions in wässeriger Lösung*, in: Z. Elektrochem. 34 (1928), 546–566. Hückel passed the following judgment on this paper in his autobiography: "What resulted was a false theory for the mobility of the hydrogen ion in water." Cf. *Ein Gelehrtenleben*, p. 121.

in the field of colloid chemistry under Donnan at the department of chemistry of University College in London. On April 12, 1928, at the age of 32, Hückel left Zurich for London with high hopes and in good spirits. He traveled alone, because his wife was expecting her first child. In a letter to Debye he reported about his departure and a few people he had met at Göttingen:

> You have probably already heard that I have meanwhile received my stipend on time after all. I found the telegram waiting for me upon my arrival in Göttingen. I am now leaving for London this evening and think that it will be a fine and stimulating period for me. [...] In Göttingen I visited Hilbert and Born. Hilbert is feeling very well again; I heard that he is even dancing again. Born was extremely nice to me, I felt I was welcomed even more kindly than ever. He told me, incidentally, that I had been on the nominating list at Rostock at his instigation.[152]

This list concerned candidates for an *Extraordinariat* in theoretical physics at the University of Rostock. The lucky recipient of the call on August 31, 1927 was the favored candidate Born's assistant, Hund, a specialist in atomic theory.[153] In any event it was certainly very advantageous for Hückel's future scientific career to be placed on such a list at Born's recommendation, even though, as will be discussed later, his life took a very different course.

When Hückel arrived in London he started going through the current research in the field of adsorption from solutions in order to compile a critical survey of the subject. During these preparations he noticed that the available experimental material "was too incomplete and inhomogenous." The theories also appeared to him to be a "mad mess."[154] Hückel soon realized that it was senseless to carry out his plan. On the side Hückel occupied himself in London "with the problems of surface potentials of pure liquids and solutions and the problems of homogenous gas reactions."[155] He considered writing a survey article on Gibbsian thermodynamics of boundary layers. But he abandoned this plan as well because the topic was, in Hückel's opinion, already "irrefutable and not new."[156]

Hückel explains further in his autobiography that Donnan always wanted to know from him "whether the statistics of the new quantum theory applied to the behavior of individual particles or to an entire set of particles."[157] Hückel unfortunately could not answer Donnan's persistent question because he had not studied the new quantum and wave mechanics. Thus Hückel became aware that he was misplaced. He

---

[152]Debye to Hückel, Zürich, 12. April 1928, MPG-Archiv, III. Abt., Rep. 19, NL Debye.

[153]Cf. Rechenberg, H.: *Die erste Göttinger Zeit und die Rostocker Professur (1919–1929)*, in: Hundert Jahre Friedrich Hund. Ein Rückblick auf das Wirken eines bedeutenden Physikers, hrsg. von Manfred Schroeder, Nachrichten der Akademie der Wissenschaften in Göttingen, mathematisch-physikalische Klasse, Vandenhoek & Ruprecht in Göttingen, 1996.

[154]*Chemiker im Gespräch: Erich Hückel*, p. 182.

[155]Original English. RAC, Box 52, Folder 805, "Report on the activities of Dr. E. Hückel during his fellowship 1928/29".

[156]Hückel, E.: *Ein Gelehrtenleben*, p. 125.

[157]Ibid.

recalled: "I realized that I had bet on the wrong horse and therefore had to readjust my plans completely."[158]

What was to be done? He had to find a way out. The decisive advice came from Debye, who happened to meet him in London. Debye referred him to Schrödinger's formulation of quantum theory as wave mechanics and "thought it would be a promising prospect to try to apply it to problems in chemistry."[159] It sounded reasonable to Hückel. He realized that the new theory was of decisive importance in atomic and molecular processes. But he was far from sufficiently prepared to apply the newly posited wave mechanics to a specific chemical problem. His first priority was to familiarize himself with the latest developments in the fields of quantum and wave mechanics along with its associated applications to chemistry.

Without the right contacts it was not such an easy matter to catch up on the latest developments. A series of lectures that Debye organized in July 1928 in Leipzig on the topic "Quantum theory and chemistry" offered a good opportunity for learning about the first applications of quantum theory to chemistry. This series, which later became known as the "Leipzig lectures" had been organized with the support of the local Ministry of Public Education in Saxony.

Donnan tried to convince the International Education Board that Hückel's new project would benefit not just from attendance at the Leipzig conference just mentioned, but also from a few weeks' stay in Göttingen among atomic theoreticians and mathematicians. He wrote to the director responsible for Europe, Wilbur Earle Tisdale of the International Education Board:

> On consultation with Professor Debye, the special subject of research selected for Dr. Hückel is *the relationship between chemical valency and the newer quantum mechanics*. I recently heard that Professor Debye was holding a Conference of Physicists at Leipzig on this special topic, beginning June 18th, and as several of the people who are eminent in this special branch of science will be present, I considered it a very important thing that Dr. Hückel should be present at this Conference. Moreover, as Dr. Hückel wished to consult various mathematicians in Göttingen concerning the special form of mathematics required for this investigation, I have agreed that after Dr. Hückel attends the Physical Conference at Leipzig, he should go to Göttingen for several weeks.[160]

So Hückel planned to study the latest findings on the links between chemical valence and quantum mechanics. Only afterwards, some months later, would he think about choosing a topic of his own. "He will be ready to start his special investigation early in September, after having consulted the best authorities on the subject," Donnan continued to Tisdale.[161] Hückel was unfortunately not permitted to make use of his London grant in his home town.[162] So he decided to delay his fellowship

---

[158] Ibid.

[159] *Chemiker im Gespräch: Erich Hückel*, p. 182.

[160] RAC, Box 52, Folder 805, Letter Donnan to Tisdale, June 18, 1928. (no emphasis in the original English).

[161] Ibid.

[162] RAC, Box 52, Folder 805, Letter Tisdale to Donnan, June 22, 1928.

by a quarter year and travel to Leipzig at his own expense, continuing on afterwards to Göttingen, where he stayed until the end of August 1928.

Around 35 scientists met at the conference in Leipzig.[163] (They included Fritz London, N.V. Sedgwick, P.A.M. Dirac, Walter Kossel and Arnold Eucken). Their presentations and debates concerned problems directly related to electrons as well as the "theory of the interactive forces between atoms and molecules" (chemistry) from the point of view of quantum mechanics.[164]

Hückel justified his decision in a letter to Debye: "I have decided to come to Leipzig because I believe that it could be of great use to my research."[165] He also informed Debye about his intention to continue on to Göttingen afterwards in order to seek closer personal ties with the pioneers of the new quantum mechanics.

At Göttingen Hückel started with a general preparation of the mathematics involved. He brushed up his skills with a close study of the standard textbook by Richard Courant and David Hilbert on *Methoden der Mathematischen Physik*. Hermann Weyl's fundamental work *Gruppentheorie und Quantenmechanik* was another of the books he took up early on. Weyl went beyond developing the basics of group theory and quantum mechanics to draw correlations between them. Besides describing Heisenberg's and Schrödinger's formalisms he depicted his own original mathematical approach to quantum mechanics. His book provided Hückel with quite a few "hard nuts to crack."[166] To get a better handle on the new quantum mechanics Hückel also scoured the original papers by Heisenberg, Schrödinger, and others. Inspired by Debye, he studied Schrödinger's papers on wave mechanics with particular intensity. They "really yielded something."[167]

At Göttingen Hückel also witnessed the birth of his first child. He later told about this fine experience:

> I still remember an emotive evening walk with Anne through the fields near Göttingen. On July 3rd [1928] we were taking another walk with Dirac and Tamm, who were also both in Göttingen, going uphill in the direction of Kaiser Wilhelm Park. There were glowworms everywhere, more than I ever saw again. Anne's labor pains started during the following day; she was taken to Dr. J. Voigt's private clinic. [...] I was allowed to be present at the birth, which started a day later, however.[168]

It was a boy. The name his parents chose was Georg Richard.[169] Life in Göttingen that summer was very harmonious for the Hückels, unsettled only by worries about whether the assistantship in Zurich was safe. It was in this regard that Erich wrote to

---

[163]Debye, P.: *Interview*.

[164]These Leipzig lectures were published a few months later as a book, in order to be accessible to a wider audience. Cf. *Leipziger Vorträge 1928, Quantentheorie und Chemie*, hrsg. von H. Falkenhagen. Hirzel Verlag, Leipzig, 1928.

[165]Debye an Hückel, Zürich, 14. Juni 1928, MPG-Archiv, III. Abt., Rep. 19, NL Debye.

[166]Hückel, E.: *Ein Gelehrtenleben*, p. 126.

[167]Ibid.

[168]Ibid., p. 127.

[169]The Hückels had four children in all. The next two boys were born in Stuttgart: Bernhard and Manfred in 1931 and 1933. The only daughter Irene was born in July 1944 in Marburg.

Scherrer shortly before his departure for London. Scherrer had meanwhile become the director of the institute of physics at the ETH. Hückel informed him about his intention to return to Zurich upon the termination of his fellowship. "He [Scherrer] wants to talk to the president about the matter as soon as possible," Hückel reported to Debye in the above-mentioned letter dated April 12, 1928 shortly before his mentor's departure from Zurich, "and thinks that he will come through with it."[170] Four months later Hückel wrote Debye from Göttingen:

> I have still not heard from Scherrer about my position in Zurich, even though I wrote him more than once since. That is why I probably can't count on being able to return to Zurich next spring at the end of my stipend and would consequently be very grateful if you think of me if you happen to hear of an assistantship for which I might come into consideration. I want to discuss this with Born in the coming days as well.[171]

At the end of August 1928 Hückel returned to England, stopping first at the Isle of Wight for a fortnight to recuperate after his taxing studies of quantum mechanics.[172] In London he returned to his studies. In a long letter Hückel reported to Debye about his recent stay in Göttingen and his plans for the future:

> I used the time in Göttingen diligently to study quantum mechanics and have made some fine progress, albeit not as much as to be able to commence my own research. You know my way of working, which is quite time consuming. I now intend to spend the winter continuing to study quantum mechanics. [...] On the other hand, I consider it a very important task of mine here to make myself familiar with as much of the research on chemistry and physical chemistry as possible.
>
> It would certainly be in the interest of the cause, I would think, if I could have my grant extended either for here or perhaps in Copenhagen – depending on the outcome of my quantum research.[173]

These lines reveal Hückel as a careful and thorough worker, publishing his work only after a lengthy gestation.

Hückel was interested in extending his fellowship, which was due to expire on July 1, 1929, because he was worried that he might find himself not only grantless but even bereft of his assistantship at the beginning of the coming summer. This risk would fall away, of course, if he received a positive signal from Zurich. "I never received a reply to any of my various inquiries to Scherrer about this and I think you will understand my not wanting to write to Scherrer again in this regard," he continued in his letter to Debye.[174]

---

[170] Debye an Hückel, Zürich, 12. April 1928, MPG-Archiv, III. Abt., Rep. 19, NL Debye.

[171] Hückel an Debye, Göttingen 5. VIII. 1928, MPG-Archiv, III. Abt., Rep. 19, NL Debye.

[172] RAC, Box 52, Folder 805, Trip by W. E. Tisdale and W. J. Robbins to England and Scotland, September 3rd to September 8th, 1928. September 5th, 1928. Rode to University College. "Huckel, E. A. Reported by student to be away on a vacation which he needed because of hard summer`s work in Germany." (original English). By nature always of fragile health, Hückel was still weak after his long illness.

[173] Debye an Hückel, Zürich, 26. September 1928, MPG-Archiv, III. Abt., Rep. 19, NL Debye.

[174] Hückel an Debye, Göttingen 5, VIII. 1928, MPG-Archiv, III. Abt., Rep. 19, NL Debye.

Hückel returned to Göttingen to see his family for a few days during the Christmas holidays at the turn of the year 1928/29. On December 27 he received a letter from Debye that surely dampened his spirits. It is worth reproducing this letter almost in full because it brings to light the precise circumstances of the state of Hückel's research and his future prospects.

> Dear Hückel,
>
> From what Sack wrote you, you can see that your situation has annoyed me very much. It is not you who are the source of this irritation, but Scherrer. I would never have thought that he would proceed as he did. Scherrer also wrote me a letter in which he argues that President Rohn is not willing to return the better position to you on the reason that he had meanwhile made it available to Pauli. Besides he (Rohn) is afraid that after a while you will lay claim to the position for physical chemistry at the Polytechnic in Zurich, and he wanted to avoid that. Just a lot of weak excuses! Scherrer continues writing whether I couldn't apply to Eucken for a position for you.
>
> This letter, which at least shows that Scherrer is not inclined to expend the least energy, annoyed me so much that I still haven't answered it. I only did not write you because I did not know, and now still do not know, how to advise you. From your letter I gather that you do not want to return to Zurich and I understand that. But it is not completely clear whether burning all your bridges to Zurich so soon would be practical instead of retaining a foothold there for the interim. For I hear that Scherrer is compelled to employ an engineer as his assistant. Under such circumstances I think he should really be glad to be able to keep you on at least for a while longer. My silence toward Scherrer at least has the advantage that I have not interfered in the least or gotten in the way of any decisions you might want to reach.
>
> We could, of course, discuss your situation, but I wonder what use such a conversation would have. I am quite at a loss about your case [...].[175]

The first dark clouds were gathering over Hückel's scientific horizons. It is remarkable that Debye regarded the reasons his colleague in Zurich had given for not continuing to engage Hückel as "weak excuses" and that he was even "quite at a loss" about his case. For a better understanding of Hückel's precarious situation, we must set it in context. Towards the end of the 1920 s a number of centers were emerging besides Munich and Göttingen, producing a generation of influential professors with a new profile in quantum theory.[176] Hückel, however, had to some extent lost touch with the momentous developments in modern theoretical physics. He was not yet familiar enough with quantum and wave mechanics to deliver original work on the subject. Consequently Hückel did not number among the group of young quantum mechanicians, many of whom were soon to receive full professorships and directorships at institutes of theoretical physics.[177]

---

[175] SBPK, Papers of Hückel, Box 6, Folder 5.13, Letter Debye to Hückel, Leipzig, 27. Dezember 1928.

[176] Cf. Eckert, M.: *Die Atomphysiker*, Chap. 4.

[177] Eckert has shown that during the winter semester 1929/30 the new focus was represented by pupils of Sommerfeld at 12 universities in Germany and in Zurich: Lenz (Hamburg), Fues (Hannover), Joos (Jena), Kossel (Kiel), Heisenberg (Leipzig), Sommerfeld (Munich), Kratzer (Münster), Ewald (Stuttgart), Landé (Tübingen), Ott (Würzburg), Pauli (ETH, Zurich), Wentzel (University of Zurich). Cf.: ibid., p. 103 and footnote 69 in Chap. 4.

I mention here a few of the more spectacular examples: Heisenberg (1901–1976), who was 5 years younger than Hückel, received the full professorship in theoretical physics at Leipzig in the fall of 1927. Two years later, on February 2, 1929, Hund joined his friend Heisenberg as the second full professor of theoretical physics there. Upon the departure of Debye to Leipzig, calls to refill his professorship in theoretical physics with a representative of the new elite were the most vociferous in the university file, as Michael Eckert has shown.[178] Only advocates of the new quantum mechanics appeared on the reduced list of candidates, with Heisenberg appearing first followed by Pauli. When Heisenberg accepted the call to Leipzig in the fall of 1927, Pauli moved up the list as the next most promising candidate. "Pauli is considered a very competent physicist, belonging to the Broglie–Heisenberg–Schrödinger school (quantum and atomic physics)," is noted in the minutes.[179] Pauli (1900–1958), 4 years younger than Hückel, received the call to the full professorship of theoretical physics at the ETH in Zurich in 1928. He requested approval of an assistantship as a condition for his acceptance, since he needed "a decent quantum man."[180] They initially wanted to transfer an assistant versed in theory from the neighboring chair for experimental physics, but Pauli refused this categorically, as Pauli explained to President Rohn of the ETH, "because I need an assistant who can work on modern atomic theory."[181] Rohn's and Scherrer's attitude toward Hückel thus seems less "weak" than it had appeared to Debye in his letter, although one cannot completely exclude a subliminal animosity toward Hückel. All of this could only have reinforced Hückel's determination to work intensely in the field of quantum mechanics.

Hückel did not give up hope and returned to London early in the new year 1929. "After my returning to London I was mainly engaged to study the physical literature on quantum theory and its chemical applications to chemical problems as well as the chemical side. In this connection I gave a course of lectures on these questions at the University College at London"[182] This was the best method by which Hückel could immerse himself in this new field.[183] Tisdale's colleague in London W. Robbins gained a better impression of Hückel by these lectures. He wrote in his notes:

> Saw him in Donnan`s office. He made a better impression on me than the last time I met him. He is evidently getting along well. Is giving a series of 4 or 5 lectures on wave mechanics for chemists, physiologists and others, omitting much of the mathematics. Over 100 came to the first lecture. The hall being overcrowded they have arranged to have the balance of the lectures given in a room seating over 200. This seems to me to be very fine evidence of the mutual exchange idea in the fellowship program.[184]

---

[178] Ibid., pp. 100–104.

[179] Eckert, M.: *Die Atomphysiker*, p. 101.

[180] Ibid., p. 103.

[181] Ibid., p. 103.

[182] Original English. RAC, Box 52, Folder 805, Report on the activities of Dr. E. Hückel during his fellowship 1928/29.

[183] Hückel, E.: *Ein Gelehrtenleben*, p. 126.

[184] Original English. RAC, Box 52, Folder 805, From Dr. W. S. Robbins, London Feb. 12, 1929.

In February 1929 Hückel visited Dirac for three weeks at the University of Cambridge. He later remarked in his autobiography, however, that these conversations "were not of substantial importance."[185] "I saw there the Cavendish Laboratory and met Sir E. Rutherford, Prof. Fowler, Prof. Kapitza, Prof. Rideal and others. Further I visited the University of Oxford, where I gave a colloquium on my mobility work and communicated especially with Prof. Sidgwick and Prof. Hinshelwood. At London I visited the National Laboratories at Teddington, the Royal Institution and the chemical Laboratories of Kings College."[186] The impression Hückel left after such visits was mixed. Ralph Fowler, for example, apparently said during a conversation with Tisdale: "Hückel is the weakest of the German group which we have had."[187]

1.9.3 Visit at Niels Bohr's Institute in Copenhagen: The Beginnings of the Quantum Theory of Double Bonding

Hückel's preparations had progressed far enough for him to feel confident about attempting his first incursion into quantum theory. He decided to spend the last three months of his fellowship (from mid-March to the end of June 1929) visiting Niels Bohr in Copenhagen. He wrote Bohr a letter in which he suggested these plans:

> I currently have a stipend of the International Education Board that runs until the 1st of July this year. Until now I have been working here in England mainly with Professor Donnan on problems in physical chemistry, but also familiarizing myself on the side with the new quantum theory by studying the literature. I would now very much like to come to Copenhagen at the end of the current term (around the middle of March) and spend the remaining quarter working on a problem in quantum mechanics.[188]

Bohr agreed in a hearty letter to Hückel as follows: "It was a pleasure for me to learn from your letter that you wish to spend the last quarter of your stipend on research in Copenhagen, and you shall be very welcome here at the institute."[189] Since its establishment on March 3, 1921, Bohr's institute in Copenhagen soon developed into a stopover for young researchers from the entire world with an interest in theoretical physics. Many wrote important papers as a result of their stays in Denmark. Bohr placed particular emphasis on international collaboration among

[185] Hückel, E.: *Ein Gelehrtenleben*, p. 129.

[186] Original English. RAC, Box 52, Folder 805, Report on the activities of Dr. E. Hückel during his fellowship 1928/29.

[187] Original English. RAC, Box 52, Folder 805, From Doktor W. E. Tisdal`s log., p. 6. Cambridg, February 4, 1930.

[188] Hückel to Bohr, London, 7. Februar 1929. NBA.

[189] RAC, Box 52, Folder 805, Letter Bohr to Hückel, Kopenhagen, 13. Februar 1929.

qualified younger researchers toward furthering the development of theoretical physics.[190]

Shortly after receiving Bohr's reply, Hückel wrote a letter to Tisdale at the Rockefeller Foundation about his intentions. The Rockefeller Foundation had taken over the fellowship program of the International Education Board after its dissolution in 1928.[191] "Since I am working on the problems concerning the application of quantum theory to chemical problems a cooperation with Professor Bohr would be very useful for my work."[192] Upon receiving the sanction of his mentor Debye a few days later, Hückel traveled to Copenhagen in early April 1929 to Bohr's institute of theoretical physics.[193]

> When I arrived in Copenhagen, an icy wind was blowing through the streets. One could still feel the cold of the past winter. But I found a nice two-room apartment with kitchen and veranda in the proximity of Copenhagen, in Taarback north of Klampenborg, very near the Baltic coast. Anne soon followed with our little Richard. [...] I usually went in town to Bohr's institute in Blegdamswej, not by streetcar from Klampenborg, but by bicycle from Taarbaek. I stayed in town at noon, usually making do with some cake; only rarely did I join the other colleagues at a restaurant because it unfortunately would have been too expensive in the long run.
>
> Near Taarbaek there were fine walks in Dürehave, a large park-like area in which red deer could frequently be sighted. There was also a fair ground with the usual popular amusements.
>
> Right near our apartment there was a small swimming pool. The first time I used it, the water was 9°C – I jumped out again as fast as I had jumped in. That was the fault of the outgoing winter.[194]

Thus Hückel commenced his work at Bohr's institute of theoretical physics in the crisp and cold idyllic atmosphere of the north. He discussed quantum mechanics and its possible applications not just with Bohr personally but also with his coworkers and guests. Since Hückel had been focusing on problems in chemistry at London, his

---

[190]Cf.: Robertson, P.: *The Early Years: The Niels Bohr Institut 1921–1930*, Akademisk Forlag, Copenhagen 1979; Pais, Abraham: *Niels Bohr`s Times*, in: *Physics, Philosophy and Polity*. Oxford University Press, 1991; Röseberg, Ulrich: *Niels Bohr: Leben und Werk eines Atomphysikers 1885–1962*. Akademie-Verlag, Berlin, 1987.

[191]For a history of the IEB see Gray, G. W.: *Education on an international scale. A history of the International Education Board 1923–1938*. Harcourt Brace, New York, 1941. Noteworthy among the current works is Siegmund-Schultze, R.: *Rockefeller and the Internationalization of Mathematics Between the Two World Wars*. Birkhäuser Verlag, Basel; Boston; Berlin, 2001, (with a comprehensive bibliography).

[192]Original English. RAC, Box 52, Folder 805, Letter Hückel to Tisdale, London, March 8th 1929.

[193]Tisdale wrote to Debye: "As the sponsor of Dr. E. Hückel, who is now a fellow of the Rockefeller Foundation, working with Professor Donnan in London, (...) I am permitted to inform you that the request has been approved, and that Dr. Hückel is hereby authorized to make the change on or about April 1st, at the convenience of himself and Prof. Bohr, and the Fellowship Committee in Natural Sciences of the Rockefeller Foundation, is prepared to pay the necessary travel expense involved" (original English). RAC, Box 52, Folder 805, Tisdale's letter to Debye, March 11th 1929.

[194]Hückel, E.: *Ein Gelehrtenleben*, p. 132.

first conversations with Bohr concerned the innovations by Heitler and London.[195] The latter had both found an explanation for the chemical bond between two hydrogen atoms in a hydrogen molecule by means of Schrödinger's wave mechanics and had approximately calculated the binding energy.[196]

A few days after Hückel's arrival in Copenhagen, the "first spring Copenhagen conference" took place, from April 8 to 12, 1929. While still preparing for this conference, Bohr wrote Pasqual Jordan:

> The plans for this conference arose from the circumstance that many physicists who had worked here before had announced their intention to visit Copenhagen during the Easter holidays. Since we can count on the attendance of Kramer and Pauli, among others, there may well be occasion for some lively and instructive discussions.[197]

Léon Rosenfeld goes down the list of participants in his "Memoirs of the first Copenhagen conference." Like Hückel, Rosenfeld was one the upcoming generation of beginners who enthusiastically followed in the footsteps of the pioneers:

> It suited Bohr's fatherly manner to be inspired by this chance gathering of visitors to call together a complete family reunion. It was not possible for all to attend but a good twenty of them followed the invitation from throughout Europe: Bohr's old friend from his Manchester period, Darwin, came from Cambridge; another close friend, one of the great masters of the classical era, Ehrenfest, came from Leyden, accompanied by his youngest pupil, Casimir; Bohr's first coworker, Kramers, came from Utrecht, who had left Copenhagen three years before to take a chair at Utrecht; Pauli came from Zurich, whose acute critical reasoning Bohr tended to call, then already, the physicist's conscience; a number of others came from Holland, Germany and Scandinavia, who each in their own special field played a part in the construction of atomic theory: Goudsmit and Kronig, Hückel, Fues, Jordan, Heitler and Nordheim, Rosseland, Holtsmark and Waller.[198]

According to Rosenfeld's account, there had been no set program. Instead Bohr took each of the participants aside to ask him what topic he would like to have discussed. A hefty debate ensued after every presentation. Over all, the need for a radical break with tradition was pointed out repeatedly, particularly by Bohr. Two years earlier the discussions at Copenhagen had revolved around Heisenberg's recently discovered uncertainty relations and how to link the conceptions of particles and waves in atomic physics. These were the most important milestones toward a physical interpretation of quantum mechanics, what came to be called the "Copenhagen interpretation."[199] During the conference Bohr emphasized that one

---

[195]Hückel, E.: *Ein Gelehrtenleben*, p. 131.

[196]Heitler, W., London, F.: *Wechselwirkung neutraler Atome und homöopolare Bindung nach der Quantenmechanik*, in: ZP 44 (1927), 455–472.

[197]Bohr's letter to Pasqual Jordan March 5, 1929. Cited from Rosenfeld, Léon: *Quantentheorie 1929: Erinnerungen an die erste Kopenhagenkonferenz, in Niels Bohr 1885–1962, Der Kopenhagener Geist in der Physik*. Friedr. Vieweg & Sohn, Braunschweig/Wiesbaden, 1985.

[198]Rosenfeld, L.: *Quantentheorie 1929: Erinnerungen an die erste Kopenhagenkonferenz*, in: *Niels Bohr 1885–1962, Der Kopenhagener Geist in der Physik*. Friedr. Vieweg & Sohn, Braunschweig/Wiesbaden, 1985, p. 297. A group photo published in this book documents Hückel among the participants. He is the unassuming figure in the third row. See the photo in Appendix II.

[199]Cf. Pais, A: *Niels Bohr`s Times*, Chap. 14.

had to be prepared to abandon the use of classical concepts. Rosenfeld described his introductory talk in his memoirs:

> He began with a few general observations that were doubtlessly designed to convey to the audience the special sense of having suddenly lost a firm footing, which is a very effective means of encouraging a receptivity toward complementary thinking. After this introductory purpose had been reached, he moved eagerly to his main topic and astounded us all (with the exception of Pauli) with the non-observability of an electron's spin. I spent the afternoon with Heitler pondering over the meagre fragments of obscure wisdom that we had managed to jot down in our notebooks.[200]

Heitler's talk about homopolar bonding on a quantum mechanical basis, which without a doubt interested Hückel the most, set Pauli in a rage. Rosenfeld's recounting of this incident illustrates very well both Pauli's temperament and the prevailing mood at the conference:

> Pauli was, as far as I can recall, quite restrained, with the exception of one spectacular occasion: Heitler's talk about homopolar bonding unexpectedly kindled his anger, because, as it turned out, he had a strong aversion to this theory. Heitler had barely finished when Pauli rushed to the blackboard in a state of agitation; striding back and forth, he began to give vent to his annoyance while Heitler sat down on a chair at the corner of the podium. "At great distances," Pauli exclaimed, "the theory is certainly wrong, because there we have Van der Waals attraction; at small distances it is, naturally, likewise completely wrong." At this point he had reached the end of the podium directly opposite Heitler's place. He turned around and went up to him pointing a piece of chalk threateningly in his direction. "And now," he thundered, "a postulate is made, appealing to the simple credibility of physicists, that claims that this model, which is wrong at large distances and wrong at small distances, is nonetheless supposed to be right in the intermediary area!" He had now come up very close to Heitler. Heitler suddenly leaned back and the back of the chair gave way with a loud snap, causing poor Heitler to fall over backwards (fortunately without hurting himself all that much). Casimir, who also remembered this incident, noticed that Gamow was the first to call out "the Pauli effect!" later adding the quip: "I do occasionally wonder whether Gamow hadn't done something to that chair beforehand."[201]

It is not known whether Hückel gave a presentation at this occasion and how he participated in the discussion. In any event in Copenhagen he experienced what it meant to move about among highly talented colleagues at the forefront of physical research in an intellectually charged atmosphere. He also realized that quantum mechanics and its applications demanded a bold spirit and probing critique of the fundamental classical concepts and required a revamping of the theoretical apparatus.

Oskar Klein, Bohr's closest Swedish colleague, had the most to do with Hückel during his stay in Copenhagen. Klein was working on the reflection of electrons according to Dirac's relativistic dynamics. Klein lent Hückel a helping hand, particularly with his problems with the mathematics of quantum mechanics, and the

---

[200]Rosenfeld, L.: *Quantentheorie 1929: Erinnerungen an die erste Kopenhagenkonferenz*, in: *Niels Bohr 1885–1962, Der Kopenhagener Geist in der Physik*. Friedr. Vieweg & Sohn, Braunschweig/Wiesbaden, 1985.

[201]Ibid., pp. 303–304.

considerably younger Hendrik Casimir initiated Hückel in the theory of permutation groups.[202]

Chemistry played an important role in Bohr's atomic theory as early as 1913.[203] His famous "trilogy" of that year[204] as well as a few of Bohr's later publications until 1923[205] focused on issues that were directly connected with problems occupying chemists: an explanation of the periodicity in the properties of the chemical elements and chemical bonding.[206] It must furthermore be noted that in 1926 Øyvind Burrau carried out the first calculations of the bonding distance, or the energy of the hydrogen molecule ion (H_2^+), on a quantum mechanical basis at Bohr's institute.[207] Nevertheless, as the historians of science Anita Kildebaek Nielsen and Helge Kragh have emphasized, by World War II, Bohr's institute had not developed into a center of the newly formed discipline of quantum chemistry.[208]

Hückel's first inquiry into quantum chemistry was nevertheless inspired by an idea of Bohr's. In his autobiography Hückel explained: "So Bohr thought one might try also to explain what was called double bonding, as are assumed to occur – or if you will, do occur – in many cases between two neighboring carbon atoms in organic chemistry."[209] Hückel took up Bohr's suggestion. Thus the quantum theoretical explanation of double bonding became his first active field of inquiry in this area. In the following years he expanded his scope stepwise, by treating other unsaturated bonds, particularly aromatic compounds. From this new research agenda his scientific *opus magnum* would eventually emerge.

---

[202] Hückel, E.: *Ein Gelehrtenleben*, p. 131.

[203] Cf. Kragh, H.: *Chemical Aspects of Bohr`s 1913 Theory*, in: Journal of Chemical Education 54 (1977), 208–210.

[204] Bohr, N.: *On the Constitution of Atoms and Molecules. (Part I)*, in: Philosophical Magazine 26 (1913), 1–25; *Part II: Systems Containing Only a Single Nucleus*, in: Philosophical Magazine 26 (1913), 476–502; *Part III: Systems Containing Several Nuclei*, in: Philosophical Magazine 26 (1913), 857–875.

[205] Bohr, N.: *On the Model of a Triatomic Hydrogen Molecule*, in: Medd. Kungl. Vetenskapsakad, Nobelinstitut, 5(28) (19219), 1–16; *Über die Serienspektra der Elemente*, in: ZP 2 (1920), 423–469; *Der Bau der Atome und die physikalischen und chemischen Eigenschaften der Elemente*, in: ZP 9 (1922), 1–67; *Röntgenspektren und periodisches System der Elemente* (mit D. Coster), in: ZP 12 (1923), 342–374.

[206] Cf. Kragh, H.: *Niels Bohr`s Second Atomic Theory*, in: Historical Studies in the Physical Science 10 (1979), 123–186; *Historiography of Electronic Valence Theory*, in: Annals of Science 40 (1983), 289–295; *Bohr`s Atomic Theory and the Chemists*, 1913–1925, in: Rivista di Storia della Scienza 2 (1985), 463–486.

[207] Burrau, ø.: *Berechnung des Energiewertes des Wasserstoffmolekel-Ions* ($H^+{}_2$), in: Det Kig. Videnskabernes Selskab Matematisk-Fysiske Meddelelser 7 (1927), 24–38.

[208] Nielsen, K, A. and Kragh, H.: *An Institute for Dollars: Physical Chemistry in Copenhagen Between the World Wars*, in: CENTAURUS, 39 (1997), 311–331. At this point I would like to thank Prof. Helge Kragh for drawing my attention to this publication and the issues involved.

[209] Hückel, E: *Ein Gelehrtenleben*, p. 131.

1.9.4 Fellowship from the Notgemeinschaft at the Leipzig Institute of Theoretical Physics

At the end of June 1929 Hückel's Rockefeller fellowship was scheduled to expire[210] and so he had to look around for alternatives. Debye also tried to find a position for him as assistant at a German university. He informed Hückel about these efforts in a letter:

> As regards a potential assistantship, I corresponded with various people but the only real prospects appear to be with *Max Wien* in Jena and in Stuttgart. The last thing I heard from Wien was that he was not sure, after all, whether he could take you in particular but he would reconsider the matter again. From Ewald in Stuttgart I finally received word a few days ago that his efforts to procure an assistantship for you there have failed. I discussed the case in detail with Laue and noticed that he is of the opinion that something really ought to be done for you in any event. If I apply for it, he is willing to arrange for a stipend for you from the Notgemeinschaft for the coming year and will push through a maximum monthly amount of *300 Reichmarks*. I am of the opinion that we should keep this option in view as the very last resort.[211]

Meanwhile the chair for theoretical physics at Rostock had become vacant. The occupant, Hund, had accepted a call to the University of Leipzig on February 2, 1929 as second *Ordinarius* of theoretical physics beside Heisenberg. Christian Füchtbauer (1877–1959), full professor of experimental physics had applied to Debye in the name of the faculty to evaluate Hückel as a possible successor of Hund. The other names Füchtbauer listed were two young physicists Albrecht Unsöld and Eugene Wigner. The former had earned his doctorate two years earlier under Arnold Sommerfeld with a thesis on wave mechanics, in which he calculated the interatomic forces in solids using quantum mechanics.[212] Wigner, a Hungarian by birth, had made significant contributions to the quantum mechanical formalism of group theory in his papers on atomic structure and molecular spectra between 1926 and 1928.[213] Füchtbauer's list of candidates reveals that the Rostock faculty intended to follow the dominant trend at other German centers of theoretical physics and appoint as Hund's successor a young member of the new generation of quantum mechanicians. By that time Hückel had just embarked on his first project in quantum theory, but the end was not yet in sight. Debye tried nonetheless to press Hückel's candidacy in a long letter to Füchtbauer. In his letter to Hückel Debye related his efforts in some detail:

---

[210] RAC, Box 52, Folder 805, Tisdale an Hückel, 30th May, 1929.

[211] SBPK, Papers of Hückel, Box 6, Folder 5.13, Letter Debye to Hückel, Leipzig, 4. März 1929.

[212] Cf.: Eckert, M. *Die Atomphysiker*, p. 144 f.

[213] About E. Wigner see Chayut, M.: *From the Periphery: The Genesis of Eugene P. Wigner`s Application of Group Theory to Quantum Mechanics*, in: Foundations of Chemistry 3 (2001), 55–78.

> I immediately replied [to Füchtbauer] by express mail with a long epistle, in which I argued that your accomplishments are greater than the other two, that a specialist quantum theoretician would be out of the question for the position in Rostock, but that you have a very good command of quantum theory as such.
>
> That is roughly what I have done thus far, leaving aside the attempts that immediately led to nothing, and I would be the one most pleased if something were to come of the Rostock professorship.[214]

In the meantime Debye had also been conducting intense negotiations with an American, Erikson, who was employed at the University of Minneapolis. It seems that Erikson was interested in Hückel because he asked Schrödinger for information about him. "Yesterday I also received the news from Schrödinger that he also had been questioned by Erikson about you; although he mentioned other names, he also recommended you very strongly," Debye informed Hückel in a letter.[215] Debye was familiar with the U.S. from various stays as guest professor. In a letter to Hückel from March 11, 1929, he described from his own experience the pros and cons of emigrating to America:

> I was once in Minneapolis as well for a short time. The university is large and evidently has quite good resources. Both Erikson and Lind are nice people, so with a bit of good will on one's own part, one could certainly arrange for a decent position for oneself there.
>
> But if you do go to America, you must be very aware from the outset that the Americans demand labor from their professors, that is teaching obligations with students. Only those who carry out this business energetically from the start will succeed. Furthermore, it is also necessary to obtain research results, if the professor in question is to have a hope of gradually improving his position. These two conditions, which have to be met simultaneously, place a heavy burden on lecturers that can only be endured with success by a very small fraction of them. The offer of $4,500. – is quite reasonable for America; it allows a modest existence. The city of Minneapolis is growing very rapidly, the environs seemed interesting to me; but you would have to be prepared for long, hard winters, because for American circumstances Minneapolis is already quite far north.[216]

No documentation about Hückel 's response has survived. Nor is it known whether Hückel received or declined an offer from the University of Minneapolis. There was another opportunity for Hückel to emigrate to America, likewise to Minnesota, as the successor of Edward Condon, who had received an appointment to Princeton.[217] In his autobiography Hückel writes: "I encountered refusals everywhere, despite all of Debye's efforts. Taking a position with Debye would not have been appropriate either. It would have destroyed my independence as

---

[214] SBPK, Papers of Hückel, Box 6, Folder 5.13, Letter Debye to Hückel, Leipzig, 4. März 1929.

[215] SBPK, Papers of Hückel, Box 6, Folder 5.13, Letter Debye to Hückel, Leipzig, 11. März 1929.

[216] Ibid.

[217] This is drawn from the following brief note by Tisdale: "HÜCKEL is to go to Minnesota next fall to take the place being made vacant by Condon`s departure for Princeton". RAC, Box 52, Folder 805, p. 39, from Doktor Tisdale`s log., January 22, 1930, Leipzig.

a scientist."[218] In any event, Hückel accepted Max von Laue's "hardly glamorous, but nonetheless [...] substantial support."[219] Through Debye's application he received from the fall of 1929 a fellowship from the Emergency Association of German Science (*Notgemeinschaft der deutschen Wissenschaft*), the forerunner of the national research association *Deutsche Forschungsgemeinschaft*. It supported research conducted at the department of theoretical physics of the University of Leipzig, where Heisenberg and Hund were full professors of theoretical physics and mathematical physics, respectively. In this period the institute of theoretical physics in Leipzig had developed into one of the internationally acclaimed centers, alongside Munich and Göttingen. It drew an influx of young physicists from throughout the world, including Edward Teller, Laszlo Tisza, Lev Landau, George Placzek, and John Slater. Their aim was to apply the new concepts and formalisms of quantum theory to numerous fields of physics and to transfer them into other disciplines as well.[220] The substance of their research applied quantum mechanics to problems in molecular physics and chemistry, leading to the emergence of new subdisciplines in theoretical physics as well as to new hybrid disciplines like quantum chemistry.[221]

In Leipzig the Hückel family found a spatious apartment on Gletscherstrasse. "It was also in the vicinity of a cemetery and a pub called 'Napoleonstein.' The funeral processions tended to stop by there after the burial of the deceased to wash down their sorrows in heady cheer," as Hückel describes it with a touch of sarcasm in his autobiography.[222]

Heisenberg was not just a dedicated scientist. He liked to entertain his colleagues as well. His coworkers and out-of-town visitors were often invited guests at his private home. Hückel and Heisenberg knew each other well since his assistantship in Göttingen, so the Hückels also benefited from his hospitality: "Anne and I were invited a few times to the Heisenbergs'. He was an excellent pianist and Anne and he played music together as well."[223]

A group of young theoreticians at the Leipzig institute was closely linked with Bohr's circle in Copenhagen as well as with Born's in Göttingen. Altogether, a strong sense of scientific ambition permeated the institute but the mood could also be generally described as cheerful and relaxed.[224] By character reserved and introverted, Hückel participated less in such recreation. He expressed himself in the following manner about this:

> I participated little in the social life at the institute. Many played ping-pong there in the evenings. I did not join often; I didn't think much of that sport, because one was constantly

---

[218] Hückel, E.: *Ein Gelehrtenleben*, p. 137.

[219] SBPK, Papers of Hückel, Box 6, Folder 5.13, Letter Debye to Hückel, Leipzig, 4. März 1929.

[220] Cf. Cassidy, D.: *Werner Heisenberg: Leben und Werk*. Spektrum Akad. Verlag, Heidelberg 2001, Chap. 14.

[221] Eckert, M.: *Die Atomphysiker*, Chap. 6.

[222] Hückel, E.: *Ein Gelehrtenleben*, p. 135.

[223] Ibid.

[224] Cf. Cassidy, D.: *Werner Heisenberg: Leben und Werk*, Chap. 14.

> having to retrieve the balls from the room's nooks and crannies. [...] For a while I made an attempt at real tennis but soon gave it up. Sports were not my thing.
>
> In general I preferred to spend my time with Anne or would stop in at the "Napoleonstein." [...] I often went in to drink a pint of beer and reflect on my research.[225]

Thus at Leipzig Hückel conducted his research on double bonding in an environment as stimulating as it had been in Copenhagen. It was a time when Heisenberg's choice young physicists were working on their doctoral or habilitation theses in quantum mechanics. Edward Teller and Rudolf Peierls were particularly noteworthy among them. As will be mentioned in the coming chapters, they both regularly had inspiring discussions with Hückel. Hückel also benefited from the occasional useful conversation with Debye when the opportunity arose. But their ties remained generally loose because Debye no longer exerted any influence on Hückel's new research.[226] Hückel received support from Hund and Heisenberg on quantum theoretical questions when, upon completion of the quantum theory of double bonding, he took up the theory of benzene.

---

[225] Hückel, E.: *Ein Gelehrtenleben*, p. 135.

[226] Hückel, E.: *Erinnerungen an Peter Debye*, p. 55.

Chapter 2
Erich Hückel's Research Agenda During the 1930s: Underpinning Organic Chemistry with Quantum Theory

2.1 The Quantum Theory of Double Bonding

The foregoing three Sections 1.9.2, 1.9.3 and 1.9.4 have shown how Hückel's interest in quantum mechanics and its application to problems in chemistry developed. Before turning our attention to a detailed history of Hückel's theory of double bonding, it seems appropriate to insert a brief sketch of classical models of double bonds in stereochemistry and their limitations. The emphasis will be on controversial aspects of the classical theory important in the development of Hückel's theory, which prompted him to develop his new ideas about double bonding.

2.1.1 Prehistory

The foundations of stereochemistry were laid by Joseph Achille Le Bel, Jacobus Hendricus van't Hoff and Johannes Wislicenus.[1] Van't Hoff in particular developed the tetrahedral model of the carbon atom. His conception of double bonding appeared first in a twelve-page brochure in Dutch in September 1874,[2] then in the following year in the now famous more comprehensive work *La Chimie dans l'Espace*.[3] He represented double bonding as two tetrahedrons of the carbon atom

---

[1]Details on the history of stereochemistry are found in: Palmer, W. G.: *A History of the Concept of Valency to 1930*. Cambridge University Press, Cambridge, 1965; Russel, C. A.: *The History of Valency*. Humanities Press Inc., New York, 1971. Vgl. auch: Kurt, H.: *Stereochemie vor van't Hoff und Le Bel*, in: Chemie in unserer Zeit 8 (1974), 129–134; Krätz, O.: *Das Portrait: Jacobus Henricus van't Hoff 1852–1911*, in: Chemie in unserer Zeit 8 (1974), 135–142; Weyer, J.: *Joseph Achille Le Bel 1847–1930*, in: Chemie in unserer Zeit 8 (1974), 143–158; Ramberg, P. J.: Chemical Structure, Spatial Arrangement. The Early History of Stereochemistry, 1874–1914. Asgate, Aldershot, 2003.

[2]van't Hoff, J. H.: Voorstel tot uitbreiding der tegenwoordig in de scheikunde gebruikte structuurformules in de ruimte. Greven, Utrecht, 1874.

[3]van't Hoff, J. H.: *La chimie dans l'espace*. Bazendijk, Rotterdam, 1875. The German translation by F. Herrmann appeared under the title *Die Lagerung der Atome im Raume* with a foreword by J. Wislicenus at the publishing house of Friedrich Vieweg and Sohn in Braunschweig, 1877.

A. Karachalios, *Erich Hückel (1896–1980)*, Boston Studies in the Philosophy of Science 283, DOI 10.1007/978-90-481-3560-8_2,

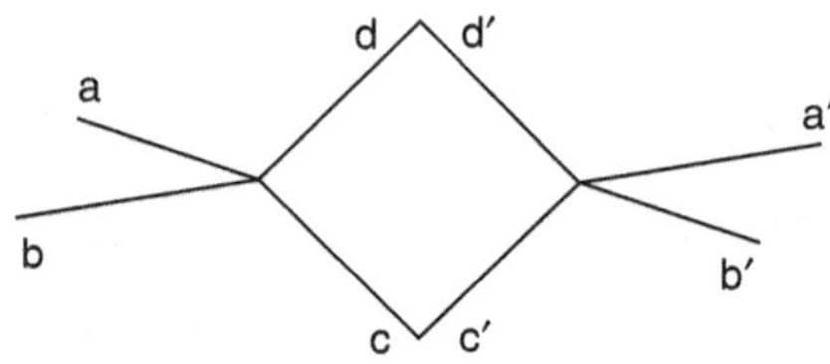

Fig. 2.1 Double bonding according to van't Hoff

joined together along one edge. The double bond in van't Hoff's pictorial model was the contact between the two tetrahedrally arranged carbon valences (Fig. 2.1).

Van't Hoff worked from the assumption that a double bond is composed of two simple bonds. His three-dimensional model illustrates how the four valences joining the central carbon atoms (cc′ dd′) appear to be bent. They form the double bond and lie on a plane perpendicular to the plane on which the other four valences are located (aa′ bb′). Any rotation around the axis of the double bond would be excluded because it would lead to a separation of the two tetrahedrons and even cause a resolution of a simple C – C bond. Another consequence of van't Hoff's model was that owing to the disposition of the plane of the substituents and the rigidity of the double bond against rotation in 1,2-disubstituent ethylene molecules, two different configurations of the isomers had to be possible that were not mirror images of themselves (Fig. 2.2) .

The substituent ethylenes are thus spatial isomers, which Wislicenus called "geometrical isomers."[4] The difference between the two models is that the positions of two given substituents are in one case on the same side of the double-bond plane and in the other case on different sides. A classic example of this isomeric form, which later came to be known as "cis-trans isomerism," is the isomeric pair fumaric acid and maleic acid. Van't Hoff's prediction in the above-mentioned paper that

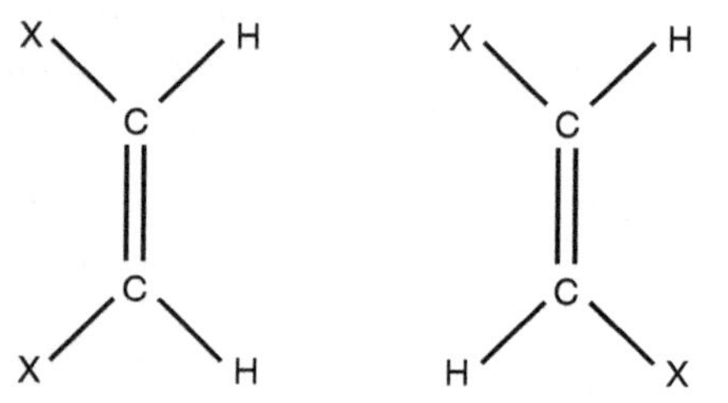

Fig. 2.2 cis geometrical isomer trans geometrical isomer

A number of other editions and revisions followed (1892, 1894, 1908). On van't Hoff's personality and work, cf. Cohen, E.: *Jacobus Henricus van't Hoff. Sein Leben und Wirken*, Leipzig, 1912.

[4]Cf.: Weyer, J.: *Hundert Jahre Stereochemie – Ein Rückblick auf die wichtigsten Entwicklungsphasen*, in: Angewandte Chemie 86 (1974), 604–611, p. 608. The terms "cis" and "trans" were introduced by Adolf Bayer in 1892. Cf. Brock, W.: *The Norton History of Chemistry*. W. W. Norton & Company, New York 1992, Chap. 7 On the Constitution and Metamorphoses of Chemical Compounds, 241–269.

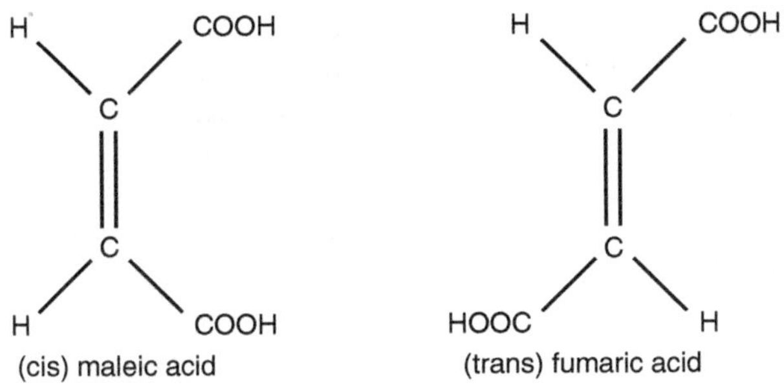

Fig. 2.3

they were isomerically arranged was confirmed by Wislicenus a few years later.[5] (Fig. 2.3)

However, van't Hoff's hypotheses raised many questions about the nature of chemical valence lines and their arrangement in space. For instance, the physics behind the tetrahedral model and the requirement of a rigid arrangement for the double bond was unknown. It was equally uncertain whether valences were really spatially directed.[6] There was no satisfactory theoretical explanation for these stereochemical phenomena. Extending van't Hoff's tetrahedral model to double bonding added another difficulty. It required the assumption of arched bonds. It meant that one could no longer speak of a regular tetrahedon model. This and other contradictions became the subject of much debate and further investigation by various chemists.[7] But the majority of them accepted a relativized tetrahedral postulate without rejecting van't Hoff's visual three-dimensional model as such.[8]

Erich Hückel's brother Walter was one of the critics of the fundamental postulates and models of stereochemistry during the 1920s who addressed the controversial problem of cyclic compounds.[9] Michael Hanack aptly described his teacher Walter

---

[5]Wislicenus, J.: Über die räumliche Anordnung der Atome in organischen Molekülen. Leipzig, 1887.

[6]The founders of stereochemistry held conflicting views about the tetrahedrally arranged valences. While van't Hoff was of the opinion that the number and direction of the valence lines must be linked to the positions of the atoms in space, Wislicenus regarded atoms as spatial structures with preferred positions of valence. He contended that the carbon atom was tetrahedrally shaped and the corresponding active zones of its valences were concentrated in the four corners of the tetrahedron. Le Bel, on the contrary, rejected the idea of spatially directed valences. He upheld the view that the spatial positions of atoms arise from an interplay between the attractive and repulsive atomic spheres. This interplay results in a particular spatial arrangement of the atoms. Cf. Weyer, J.: *Hundert Jahre Stereochemie*, (with references to the original papers).

[7]Weyer, J.: Hundert Jahre Stereochemie – Ein Rückblick auf die wichtigsten Entwicklungsphasen, in: Angewandte Chemie 86 (1974), 604–611.

[8]Weißbach, H.: *Strukturdenken in der organischen Chemie*. VEB Deutscher Verlag der Wissenschaften, Berlin, 1971.

[9]Hückel, W.: *Der gegenwärtige Stand der Spannungstheorie*, in: Fortschr. Chem. Physik u. Physikal. Chem. 19 (1928), 1–36. In this survey article Walter outlined the advances made in strain

Hückel as striking out in a new direction already in the early 1920s as compared to other more classically imbued organic chemists. When Walter published his first papers, organic chemistry was primarily a preparative chemistry and partly analytic. This meant that the goal of an organic chemist was in the first place to figure out the structure of natural substances and then to synthesize them in the laboratory. Walter, by contrast, used experiments mainly as an instrument for explaining controversial theoretical issues.[10]

At Easter 1921 Walter, as Adolf Windaus's assistant, started to investigate "the question of the existence of two stereoisomeric decahydro-naphthalines."[11] (Fig. 2.4). This new research project arose out of discussions with Windaus, one of the most prolific organic chemists of the day. They considered the limited applicability of van't Hoff's tetrahedal model in rendering the spatial relations of organic molecules. Looking back on these discussions Walter recounted in his memoirs:

> Thus emerged out of these conversations between Windaus and me, entirely of its own accord, the realization that this question was of fundamental importance in conceiving models and that an experimental analysis was therefore worthwhile. For me the existence of stereoisomeric deca[hydro]naphthalines was at the same time a somewhat more far-reaching problem, for, it appeared to me to be a subordinate task within the more general problem of the extent to which van't Hoff's carbon tetrahedrons were useful in the construction of models of the molecule.[12]

At the close of the 19th century Hans Sachse had calculated from van't Hoff's tetrahedral model that two nonplanar (bent) strain-free ring models of cyclohexane were possible. They are referred to as the chair and boat (or bathtub) configurations (Fig. 2.5). In addition, Sachse postulated corresponding nonplanar structures for larger rings.[13] So Sachse's model demonstrated that rings of up to five carbon

Fig. 2.4 The two stereoisomeric decahydro-naphthalines

theory, questioning in particular the theoretical considerations of his English colleagues Jocelyn Thorpe and Christopher Ingold. Walter reminisced in his memoirs in this regard: "Besides the fact that this paper signified for me a final closure to a debate I had long been mentally engaged in with the English scientists – subsequent research precisely by English scientists have since proven the correctness of my counterarguments – this paper was also valuable to me because in the process of writing it I realized for the first time the proper significance of the idea of the spatial extension of a molecule or an atom and recognized the limitations of applying this mechanical concept in chemistry just as in physics." Cf. Hückel, W.: *Memoiren*, p. 257.

[10] Hanack, M.: *Das wissenschaftliche Werk von Walter Hückel*, in: CB 113 (1980), V–XVII.

[11] Hückel, W.: *Memoiren*, p. 201.

[12] Ibid.: p. 202.

[13] Sachse, H.: *Ueber die geometrischen Isomerien der Hexamethylenderivate*, in: B 23 (1890), 1363–1370; *Über die Konfigurationen der Polymethylenringe*, in: ZPC 10 (1892), 203–241.

Fig. 2.5

members lie on one plane,[14] whereas strain-free rings of six and more carbon atoms are "spatially" (or three-dimensionally) structured. It also followed from Sachse's theoretical considerations that Adolf Baeyer's strain theory was not correct, which had regarded six-membered rings and all other rings as planar. Moreover, it was theoretically possible to construct strain-free rings that retain the tetrahedral angle.

In 1918 Ernst Mohr returned to Sachse's hypothesis and extended it to bicyclic systems of six-membered rings. He predicted that decaline (decahydro-naphthaline) ought to have both strain-free cis and trans forms if the planar form of cyclohexane were abandoned.[15] But illness prevented Mohr from confirming their existence experimentally. This became Walter's new task. In March 1923 he succeeded in isolating the two forms of cis and trans decaline.[16] Thus by experiment Walter Hückel refuted a conception that had long been the subject of controversy. Ring-shaped paraffins, specifically cyclohexane, were of a planar structure and, consequently, if a trans form of decaline existed at all, it was unstable.

Walter Hückel applied the problem of the extent to which van't Hoff's tetrahedral model was suited toward the construction of molecular models to saturated compounds (with no double bonds in the ring) as well as to unsaturated organic compounds (with double bonds in the ring). He continued reporting in his memoirs: "It was clear to me from the outset, however, that in reproducing the double bonding van't Hoff's requirement that the tetrahedrons be positioned next to each other along one edge brings the centers of mass of the doubly bonded atoms far too close together."[17]

No doubt the Hückel brothers frequently discussed such considerations and other critical points about the basic stereochemical postulates and models. Through his brother Walter, Erich also had the opportunity of keeping abreast of the current research in organic chemistry and stereochemistry in particular. In an interview Erich later mentioned the limitations of the tetrahedral model and the conceptions that chemists had of double bonding prior to the age of quantum mechanics:

---

[14]From today's point of view, this assumption is valid only on statistical average, i.e., the molecule oscillates in the form of envelope configurations of cyclopentane.

[15]Mohr, E.: *Die Baeyersche Spannungstheorie und die Struktur des Diamanten*, in: Journal für Praktische Chemie (Neue Folge) [2] 98 (1918), 315–353; *Zur Theorie der cis-trans-Isomerie des Dekahydro-naphtalins*, in: B 55 (1922) 230–231.

[16]Hückel, W.: *Zur Stereochemie bicyclischer Ringsysteme I. – Die Stereoisomerie des Dekahydro-naphthalins und seiner Derivate*, in: Liebig's Annalen der Chemie 441 (1925), 1–48; *Die raumisomeren Formen des Dekahydronaphthalins*, in B 58 (1925), 1449–1452.

[17]Hückel, W.: *Memoiren*, p. 253.

There actually was no theory [of double bonding] at all. What concepts did chemists have in organic chemistry? They had the concept of valency ["Wertigkeit"], of directed valences – the tetrahedral model of carbon – and for double bonding they had a model in which two tetrahedrons were connected along one edge. But that caused difficulties because the distances were wrong and couldn't be brought into agreement with the other relations. [...] It was not theoretically supported, it was just an imprecise picture. Added to that were the ideas about valence saturation, Thiele's partial valences and the like. But those were less theories than attempts at making order out of the multifarious observations.[18]

2.1.2 The Beginnings of Hückel's Quantum-Theoretical Interpretation of Double Bonding

As already mentioned, it was only in Copenhagen that Hückel commenced his new research project on a quantum-mechanical analysis of double bonding. The report to the Rockefeller Foundation on his initial research results by the end of his fellowship on 28 June 1929 reveals the first traces of his new conceptual endeavor in this direction:

There I treated especially the problem of the so called rigid valency bounds [sic!], as it is for instance to be seen from the existence of isomerism of fumar and malein acid. It could be shown not only from the chemical facts but also from theoretical considerations, that the rigidity of the double bounds in these [an]d similar compounds cannot be explained by an ordinary interaction between the substituents, but must be due to the properties of the double valency bound itself; a fact, which until the[n] had not yet been considered in the interpretation of valency by the new quantum mechanics. The great difficulties which the theoretical treatment of this problem involves have not yet been completely overcome. Notwithstanding th[is] aspect for the solution seems to be hopeful.[19]

In Copenhagen Hückel identified the problem to be treated and even ran up against theoretical difficulties. They concerned the interpretation of cis and trans isomerism of alkene derivatives, and specifically, the rigidity of double bonds against rotation. The result of his investigations located the cause not in effects of the substituents but in the nature of the double bond itself. Consequently the electron structure of double bonds had to be examined. Hückel consulted the new quantum mechanics in his search for an answer to the problem of isomerism, which could not otherwise be settled definitely by standard methods in chemistry.

Under what conditions did Hückel's concept of double bonding develop? What methods did he use? How did his quantum-theoretical approach to double bonding take shape?

First of all, it is clear that Hückel could not have tackled the problem at hand without being versed in organic chemistry and stereochemistry in particular. The

---

[18] *Chemiker im Gespräch: Erich Hückel*, in: Chemie in unserer Zeit 4(1979), 180–187, p. 182.

[19] (Original English) RAC, Box 52, Folder 805, Report on the activities of Dr. E. Hückel during his fellowship 1928/1929.

scientific exchanges he had with his brother Walter a few years before as well as his research in London under Donnan both played a decisive role.[20]

Erich had been in close contact with his brother Walter since his analyses in colloidal chemistry. Walter wrote in his memoirs in this regard:

> In 1925 my brothers got married: my brother Erich to Annemarie Zsigmondy, my brother Rudi to Ruth König, the daughter of a captain in the cavalry who had died in the war. The marriage of my brother Rudi had the consequence that I met him more seldomly than before and for a time lost sight of his scientific development. But I always continued to exchange scientific ideas with my brother Erich, whose scientific research was more closely related to mine.[21]

This contact between the elder Hückel brothers was not restricted to the exchange of ideas. It developed into a close collaboration that bore fruit a few years later in the form of Walter's two-volume work on theoretical foundations of organic chemistry.[22] As Erich related in his autobiography, the cooperation with his brother developed at the time of their father's death.[23] "As our father's corpse was being carried out of the house, my brother Walter and I were laboring over a section of his book: *Theoretische Grundlagen der organischen Chemie*, to help us get over the difficult hour. We had admired and loved our father above all else."[24]

At Leipzig Hückel continued his research on double bonding. End of December 1929 it was finished and appeared at the beginning of the following year in the journal *Zeitschrift für Physik* under the title "On the quantum theory of double bonding."[25] According to the acknowledgments Hückel had engaged in "valuable discussions about the quantum-theoretical part" of his paper with a student of Sommerfeld, Edward Teller. At the time Teller, who was twelve years younger than Hückel, was working on his thesis under Heisenberg on a quantum-theoretical topic concerning the hydrogen molecule ion.[26] Erich also acknowledged his brother Walter for discussions they had had about data and problems in chemistry.

---

[20]"My main intention for the time of my fellowship was to apply the new quantum mechanics (wave mechanics) to chemical problems. For this purpose the cooperating with Professor Donnan and his coworkers was of great use for me as to the chemical side of the problems," Hückel wrote in his report to the Rockefeller Foundation (original English). RAC, Box 52, Folder 805, Report on the activities of Dr. Hückel during his fellowship 1928/29, July 26th 1929.

[21]Hückel, W.: *Memoiren*, pp. 278–279.

[22]Hückel, W.: *Theoretische Grundlage der Organischen Chemie*. Erste Auflage, Band I, II Akademische Verlagsgesellschaft m. b. H., Leipzig, 1931.

[23]Hückel's father died on 21 August 1927.

[24]Hückel, E.: *Ein Gelehrtenleben*, p. 121.

[25]Hückel, E.: *Zur Quantentheorie der Doppelbindung*, in: ZP 60 (1930), 423–456.

[26]Upon completing his doctorate in spring 1930 Teller became assistant at Göttingen and collaborated with James Franck, Walter Heitler and Lothar Nordheim. He made a name for himself as an expert in applying quantum mechanics to problems in molecular physics. After 1933 he was forced to leave Germany because of his Jewish origins. After a brief time first in London at Donnan's institute and then in Copenhagen with Bohr, he moved to the USA in the fall of 1935. There he participated in the American atomic bomb project. Cf., e.g., Blumberg, A., Owens, G.: *Energy and Conflict. The Life and Times of Edward Teller*. London 1976; Wigner, E.: *An Appreciation on the*

Erich Hückel's paper subjected the fundamental concept in organic chemistry of tetrahedrally arranged valences (valence forces) to critical scrutiny and concretely defined the problem to be treated:

> Chemists, however, especially organic chemists, are inclined to attach more to the concept of valence than the mere valency of atoms. They want to ascribe a certain real existence to their valence lines such that, particularly in the chemistry of carbon, not just the number of valence lines but also their orientation in space become meaningful. There are indeed strong arguments in favor of this interpretation, to which as a consequence probably most organic chemists adhere despite various objections raised, particularly of late.
>
> Yet it is not the intention of this paper to treat this question in general. Rather only a special case will be examined that may contribute toward its resolution. This case concerns what chemists call "the rigidity of double bonding" against rotation.[27]

It is noteworthy that Hückel did not immediately set out in search of a quantum-mechanical explanation for the tetrahedral arrangement of valences. He focused on the resistance of the C = C double bond to rotation, which organic chemists had known about since the second half of the 19th century. Let us examine more closely how Hückel came to identify this specific research topic.

In the first part of his paper Hückel summarized the findings in chemistry and physics regarding the behavior of the C = C and the C = N double bonds. It contradicts van't Hoff's tetrahedral model and cannot be explained by basic stereochemical principles. He mentioned, for instance, that the cis and trans forms can convert into each other under the influence of certain catalysts. Under specific conditions and in the catalytic presence of bromine or a mercury salt, maleic acid can change into fumaric acid.[28] Such a conversion can also take place without a catalyst upon exposure to ultraviolet light.[29]

60th Birthday of Edward Teller, in: H. Mark, Fernbach (Ed.): *Properties of Matter Under Unusual Conditions. In Honor of Edward Teller's 60th Birthday*. New York 1969, 1–6.

[27] Hückel, E.: *Zur Quantentheorie der Doppelbindung*, p. 423.

[28] E. Hückel was thus supporting his arguments on the latest developments in organic chemistry. In 1928 F. Wachholtz published his research results on the above-mentioned topic in the *Zeitschrift für physikalische Chemie*. Cf. Wachholtz, F.: *Die Umwandlung von Malein- in Fumarsäure durch Bromatome*, in: ZPC 135 (1928), 147–181.

[29] Regarding the photochemical conversion of fumaric and maleic acids, E. Hückel referred to E. Warburg's communication in the *Sitzungsberichte* of the Prussian Acad. Science, 1919, p. 964, no. 50. His support regarding the conversion of substituent ethylenes in general was the papers R. Stoermer had published with his collaborators between 1905 and 1922. Cf. Stoermer, R. und Simon, M.: *Ueber geometrisch isomere Derivate des Diphenyläthylens und deren Configuration*, in: Liebig's Annalen der Chemie 342 (1905), 1–13; Stoermer, R.: *Über die Umlagerung stabiler stereoisomerer Äthylenkörper in labile durch ultraviolettes Licht (I)*, in: B 42 (1909), 4865–4871; *Über die Umlagerung stabiler stereoisomerer Äthylenkörper in labile durch ultraviolettes Licht (II)*, in: B 44 (1911), 637–668; Stoermer, R. und Heymann, P.: *Die Bestimmung der Konfiguration der stereoisomeren Zimtsäuren*, B 45 (1912), 3099-3104; *Hrn. Liebermann zur Erwiderung. – Zur Kenntnis der halogenierten Zimtsäuren und deren Verhalten im ultravioletten Licht*, in: B 46 (1913), 1249–1266.

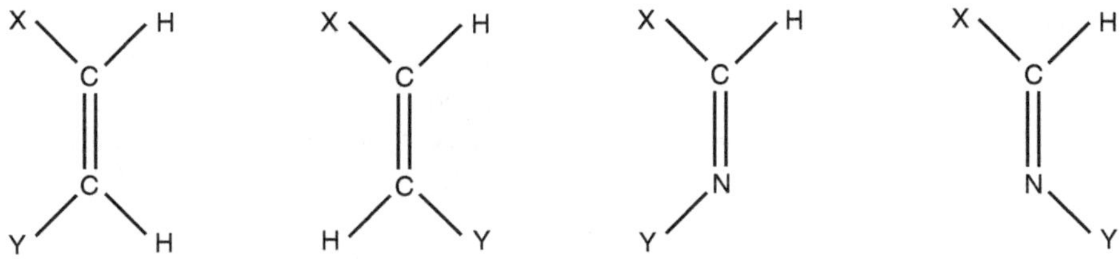

Fig. 2.6 Cis and trans isomerism and syn- and anti-isomerism

The C = N double bond is an isomeric phenomenon analogous to C = C. Toward the end of the 19th century Arthur Hantzsch (1857–1935) and Alfred Werner (1866–1919) applied Le Bel's and van't Hoff's tetrahedral model and stereochemical postulates concerning the carbon atom to the trivalent nitrogen atom.[30] They formally substituted a trivalent ["dreiwertige"] methine group (C–H) in geometrical isomeric ethylenes with a trivalent nitrogen atom (N) and transferred the stereochemical model for C = C double-bonded isomers to compounds with C = N double bonds (Fig. 2.6).

In this case one does not refer to cis and trans but syn- and anti-isomerism. As Hückel mentions in his paper, an interconversion can thus take place just as well in C = N bonding as in C = C bonding at raised temperature or in the presence of a catalyst or as an effect of ultraviolet light.[31] Hückel's analysis of the facts revealed that the cause of the behavior of C = C and C = N double bonds, inexplicable according to van't Hoff's model, could only be sought in the nature of double bonding.[32] A new theory of double bonding was needed. The new theory would have to confirm observations on ethylene isomerism just as well as van't Hoff's. It would thus have to be able to explain the nonrotatability of double bonds and the planar positionings of the substituents. These points are emphasized in the introduction to Hückel's article of 1930.

2.1.3 Excursion: Hückel's Physicochemical Experiments on the Stability of Cis and Trans Isomers

Hückel first tried to understand the stability of double bonding according to the classical foundations of physics. He considered explaining the stability of cis and

---

[30] Cf. Kauffmann, G. B.: *Alfred Werner – Founder of Coordination Chemistry*. Berlin, Heidelberg, New York, 1966; *Foundation of Nitrogen Stereochemistry, Alfred Werner's Inaugural Dissertation*, in: Journal of Chemical Education 43 (1966), 155–165; *The Stereochemistry of Trivalent Nitrogen Compounds: Alfred Werner and the Controversy Over the Structure of Oximes*, in: Ambix, 19 (1972), 129–144; Stocklöv, J.: *Arthur Hantzsch: Wegbereiter der physikalischen organischen Chemie*, Dissertation, Halle (Saale) 1996.

[31] Hückel, E.: *Zur Quantentheorie der Doppelbindung*, p. 429.

[32] Ibid., pp. 433–434.

trans forms by a special interaction between the substituents, for example, by dipole interaction. He also tried to attribute the stability of the two configurations to a particularly potent polarizability of the double bond.[33]

Hückel's exchanges with his brother Walter were important in this quest. He also attended a week-long conference organized by Debye in summer 1929. It was the second series of talks known as the "Leipzig lectures,"[34] this time covering the topic "Dipole moment and chemical structure".[35]

Walter Hückel also participated. Since accepting a call as regular extraordinary professor of organic chemistry at Freiburg (Breisgau) in 1927, attendance at such conferences was crucial to his future career in science. These Leipzig lectures are mentioned in his memoirs:

> The various conferences that I attended during the Freiburg period were significant for my scientific development. This was particularly true of the dipole conference in Leipzig that Debye organized in June 1929. Because my brother Erich was in Leipzig, I could discuss scientific matters through with him as well.

By the 1920s chemistry had advanced to the point that dipole measurements could be used to test the applicability of stereochemical conceptions of molecular structure.[36] At the beginning of 1929 Walter had just demonstrated in an article entitled 'Molecular structure and dipole moment' that the occurrence of dipole moments in some compounds of the type CR_4 did not conflict with van't Hoff's postulated tetrahedral arrangement.[37] (Fig. 2.7)

The article questioned the stereochemical views advocated by Ludwig Ebert and H. von Hartel. They contended that many methane derivatives of the type CR_4

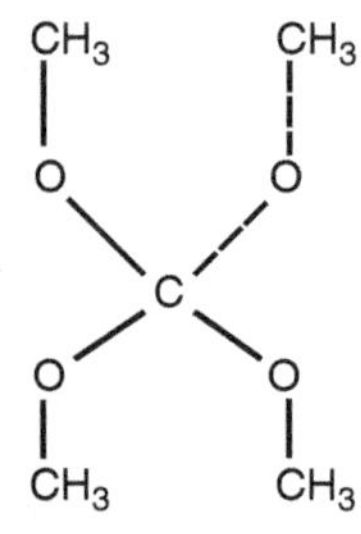

Fig. 2.7 Model of the compound $C(OCH_3)_4$

[33]Ibid., 434–436.

[34]The first Leipzig lectures under the heading "Quantum theory and chemistry" had taken place that summer.

[35]*Leipziger Vorträge 1929, Dipolmoment und chemische Struktur*, Hrg. Von P. Debye, Verlag von S. Hirzel, Leipzig, 1929.

[36]Debye, P.: *Polare Molekeln*. Verlag von S. Hirzel, Leipzig, 1929. Debye's monograph was a German translation of an English work *Polar Molecules*, published by Chemical Catalogue Company, New York. The German translation was prepared by Debye's assistant Sänger in Zurich. Debye subsequently revised it and added a chapter on polarity and its relation to chemical structure. It was the first survey of the electrical properties of molecules.

[37]Hückel, W.: *Molekülbau und Dipolmoment*, in: ZPC (B) 2 (1929), 451–457.

exhibiting a strong dipole moment did not conform with the prevailing stereochemical theory of van't Hoff. This meant that one would have to replace the presumed tetrahedral structure with a pyramid for such molecules.[38] Thus a controversy arose between Walter Hückel and Ebert. During the Leipzig lectures, to which Debye had invited both parties, there were some heated debates about the issue. Walter Hückel recalled in his memoirs:

> Debye had asked me to give a presentation at this conference, for which I chose the topic: dipole moment and reaction rate. [...] Lively discussions arose among the select participants of this conference in which, among other things, the principle of free rotation and its implications on the dipole moment played prominently. Initially there were two opposing points of view, on the one hand, the one upheld by L. Ebert and I. Estermann, and on the other, the one defended by K. L. Wolf and me. Today the latter view on the positional angle of the substituents and free rotatability, which evidently cannot be regarded as absolutely unimpeded, has become so integrated in the common knowledge that it is hard to even put oneself in Ebert's former frame of mind, who himself changed his opinion during the conference.[39]

The relations between the magnitude of the electric dipole moment and the required configuration of cis-trans-isomers such as dihalogen ethylenes were also discussed at these Leipzig lectures.[40] This was the problem I. Estermann was working on. In his talk on "Dipole moment and molecular structure" he demonstrated the possibility of using measurements of the dipole moment to distinguish the configurations of cis and trans compounds.[41]

Thus from conversations with Walter Hückel and ideas presented at the Leipzig lectures, Erich Hückel gathered important cues leading to his assumption that the dipole interactions between substituents cause the stability of double bonds against rotation. First Hückel tried on the basis of classical physics to estimate the

---

[38]Ebert, L., Eisenschnitz, R. und Hartel, H.: *Über die elektrische Symmetrie des Molekülbaus von Methanabkömmlingen*, in: ZPC (B) 1 (1928), 94–114. They expressed their view as follows: "In this sense the interpretation of our measurements is probably the conclusion that the structure of the molecules Ca_4 manifest a greater diversity than the models of classical stereochemistry were able to predict; specifically, that forms occur possessing a strong dipole moment which therefore, if one is to retain the equivalence of the four a groups, would according to Weissenberg have to have the symmetry of a pyramid." (Loc. cit., p. 114). The following compounds were at issue: $C(OCH_3)_4$ and $C(OC_2H_5)_4$ (with small dipole moments); $C(COOCH_3)_4$ and $C(COOC_2H_5)_4$ as well as $C(CH_2OOCCH_3)_4$, $C(CH_2ONO_2)_4$, $C(CH_2OH)_4$ (with considerable moments).

In 1926 K. Weissenberg had deduced purely on the basis of symmetry theory that a pyramid was possible besides the tetrahedal arrangement as the structure of compounds of the type Ca_4. A summary of Weissenberg's theoretical considerations is provided in the paper by Friedrich Richter: *Das Verhältnis der klassischen Stereochemie zu den neueren Arbeiten K. Weissenbergs*, in: Naturwissenschaften, 14 (1926), 889–895.

[39]Hückel, W.: *Memoiren*, 304.

[40]To prevent confusion about the terminology used, I specify here the historical definitions. In the original papers, the term "configuration" was understood as the spatial arrangement of atoms in a molecule; "constitution" only meant the distribution of bonds (valence lines) without considering their three-dimensionality.

[41]Cf. Estermann, I.: *Dipolmoment und Molekülbau*, in: *Leipziger Vorträge 1929, Dipolmoment und chemische Struktur*, Hrg. von P. Debye, Verlag von S. Hirzel, Leipzig, 1929, p. 36.

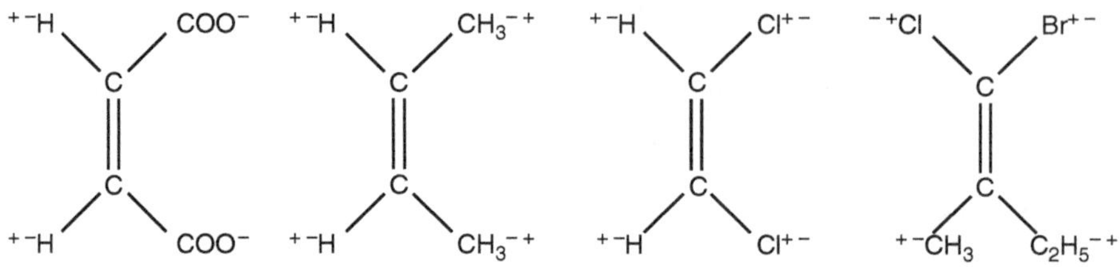

Fig. 2.8

electrostatic forces arising from the corresponding distribution of charges among indirectly linked spatially determined substituents. He analysed the following four compounds and their isomers (Fig. 2.8)

The first compound is a doubly negatively charged ion that is stable in both the cis and trans forms.[42] It was known from measurements of the dipole moments of the other three compounds that the groups C–H, C–CH_3 and C–C_2H_5 carry a dipole whose positive end is directed outwards. For the groups C–Cl and C–Br the externally directed ends of the dipoles were negatively charged. This is expressed in the formulas by the additional + – and – + signs.[43] Hückel's paper concluded that in the first case, because of the ionic repulsions, only the trans form should correspond to a stable configuration of low energy. In the second case, too, the classical explanatory pattern contradicted experimental observation: according to the given assumptions, both isomeric configurations of the butylene groups would not be stable because the planes of the two substituents would have to be angularly positioned. But Hückel knew that, in fact, only planar butylene molecules exist. The third compound was similar to the first. In the fourth instance the dipole formulas do not explain the energy differences between the isomeric configurations. Finally Hückel tried to attribute the stability of both positions to the polarizability of the double bond. His estimates revealed that even when taking the polarizability into account it was not possible to explain the observed stability of the cis and trans isomers.

Hückel thus realized that neither van't Hoff's visual representations in stereochemistry nor models used in classical physics were able to explain the experimental findings. The observed stability of the cis and trans isomers and their interconversion under specific conditions, particularly under the influence of ultraviolet radiation, remained anomalous. At the close of the first part of his paper he underscored the necessity for an interpretation according to quantum theory:

> All in all, these considerations show that we may consider it impossible to explain on a classical basis the active forces resulting in the stability of double bonds against rotation. The fact that isomeric conversion occurs under the influence of ultraviolet light is another

[42] Both disodic fumarate and disodic maleinate have measurable second constants of dissociation. Cf., e.g., Hückel, E.: *Zur Quantenteorie der Doppelbindung*, 434.

[43] For historical reasons we use Hückel's original designations.

supporting argument for seeking the cause of this double-bond behavior in the electronic structure of double bonding.[44]

Hückel's justification for the necessity of a quantum-theoretical interpretation of the stability of double bonding was more than a simple critique of the fundamental stereochemical postulates and models. He pointed out conflicting experience. One of the decisive elements of his argument was the peculiar way in which double bonds absorb light, that is, the photochemical interconvertibility between the cis and trans forms. This suggested that the cause of the rotational rigidity lay in the electronic double-bond structure itself. In Hückel's opinion, the double-bond electrons must have a characteristic arrangement that has a mutually stabilizing effect on the atoms lying on a directly adjacent plane.[45] To solve this puzzle Hückel resorted to the laws and methods of quantum mechanics. The main task thus to be accomplished was to explain the characteristic stabilizing distribution of charges among the electrons in a double bond.

2.1.4 The Quantum-Mechanical Interpretation of Double Bonding

The second part of Hückel's paper constituted the quantum-mechanical treatment of double bonding between carbon atoms. Starting from the electronic structure of the oxygen molecule (O=O) he progressed via formaldehyde ($O=CH_2$) to ethylene ($CH_2=CH_2$). Hückel first considered the O_2 molecule as a prototype. Because the O nucleus has two more nuclear charges than the C nucleus, his next step was to substitute the O nucleus with a C nucleus by removing two H nucleii (protons), which are then linked with the resulting C nucleus to yield formaldehyde ($O=CH_2$). Next, by a similar consideration concerning the oxygen molecule's second nucleus, Hückel arrived at ethylene ($CH_2=CH_2$). Where did Hückel get this original idea?

His brother Walter had mentioned in his paper, 'Contributions to the conception of homopolar bonding of atoms' from 1921[46] that some organic compounds can be reduced to "determinable prototypes" like F_2, O_2, N_2 with "stable electron arrangements" on the basis of the number of covalent carbon and hydrogen electrons. Walter Hückel had been inspired by Born's and Landé's model of the chlorine molecule as two electron cubes. With one electron assigned to each of the eight edges per cube, two shared electrons remained.[47] Walter observed that each atom no longer had an integral electron configuration of its own in the cube model by Born and Landé.

---

[44] Hückel, E.: *Zur Quantentheorie der Doppelbindung*, p. 437.

[45] Cf. *Chemiker im Gespräch: Erich Hückel*, p. 183.

[46] See the end of Section 1.6 of this work. I point out again that Walter acknowledged his brother Erich specifically as an advisor.

[47] Inspired by Kossel's theory, Born and Landé attempted on understand the crystal lattice of alkali halides between 1918 and 1920. They concluded that a satisfactory agreement between their calculations and experiment is reached if one assumes that the eight electrons move in such a way that at any point in time they can be regarded as positioned at opposing corners of a cube. Cf.: Born, M. und Landé, A.: *Über die Berechnung der Kompressibilität regulärer Kristalle*

Instead the electrons of the two atoms all contributed at the same time. That is how Walter Hückel arrived at the idea that other compounds that were not purely heteropolar also existed. Some of the electrons were shared and could not be assigned definitely to a specific atom of the molecule. I quote here a few of Walter Hückel's considerations that reveal the source of Erich Hückel's conceptual scheme quite clearly. Walter reduced a few organic compounds to the prototypes F_2, O_2 and N_2 in the following way:

> If *a hydrogen atom is removed* from CH_4, then there remains CH_3, methyl. It contains seven positive nuclear charges of which four originate from C and three from H, seven opposing C and H shared electrons as in the F atom. (The two inner electrons of the C atom are disregarded here; their charges are compensated by two positive nuclear charges.) The arrangement of these seven electrons is similarly asymmetric and unstable as in the flourine atom and a joining of two methyls into a double molecule with a stable and symmetric electron arrangement is therefore possible here as well.
>
> Analogous to binding with a second methyl, a halogen atom can also form a homopolar bond whose function is thus not to attract an electron unto itself but to fuse with the methyl, as it would with a second halogen atom to form a molecule.
>
> *Just as partial methyl has seven electrons, like the F atom, so also the radical methylene,* CH_2*, which has six electrons like O. It can form a double molecule with a similar electron arrangement as molecular oxygen,* $CH_2{=}CH_2$*; moreover, it can form the compound* $CH_2{=}O$. The great dissimilarity between these compounds and O_2 ought not be a surprise; the unsaturated character of the carbon–carbon and carbon–oxygen double bonds does not return in the oxygen–oxygen double bond because it cannot form stable low-energy bonds by the addition of H_2, for instance, as is the case with C = C and C=O bonding. [...]
>
> Methine, CH, with five electrons, is comparable to the nitrogen atom. As a double molecule it forms HC≡CH (like N≡N) acetylene; joined to N, the result is HC≡N, a form of hydrogen cyanide, which reminds one of the electron arrangement of N_2.[48]

Thus we see a clear parallel between Walter's basic ideas and Erich's quantum-mechanical treatment of ethylene, which is derived from a progression of electron configurations for the oxygen molecule and formaldehyde as the intermediary.

In September 1929, shortly before the appearance of Hückel's paper, the British physicist John E. Lennard-Jones employed the molecular orbital method (MO) to examine the electron structure of the oxygen molecule according to quantum theory.[49] For the oxygen molecule (O_2) in its ground state he indicated the following arrangement of electrons:

---

aus der Gittertheorie, in: Verh. Dtsch. Physik Ges. 20 (1918), 210–216; Born, M: *Über kubische Atommodelle*, in: Verh. Dtsch. Physik Ges. 20 (1918), 230–239; *Dynamik der räumlichen Atomstruktur*, in: ZP 2 (1920), 83–86; *Bemerkungen über die Grösse der Atome*, in: ZP 2 (1920), 87–89; *Würfelatome, periodisches System und Molekülbildung*, in: ZP 2 (1920), 380–404; Landé, A.: *Dynamik der räumlichen Atomstruktur*, in: Verh. Dtsch. Physik Ges. 21 (1919), 2–12, 644–662.

48 Hückel, W.: *Beiträge zur Auffassung der homöopolaren Atombindung*, p. 307. (Emphasis mine.)

49 Lennard-Jones, J. E.: *The Electronic Structure of Some Diatomic Molecules*, in: Transactions of the Faraday Society 25 (1929), 668–686.

$$(1s)^2\,(1s)^2\,(2s)^2\,(2s)^2\,(2p_+)^2\,(2p_-)^2\,(2p\sigma)^2\,\{2p\pi_+,2p\pi_-\}$$

There were two other theoretical MO descriptions of the electron states of a diatomic molecule besides the one introduced by Lennard-Jones: the currently much more familiar descriptions by Hund and Mulliken. Hückel pointed out in his paper that he found Lennard-Jones's formulation preferable, however, because it was better suited to representing the states electrons of a diatomic molecule undergo during nuclear decomposition.[50] This formulation fit better into Hückel's theoretical framework of quantum-mechanically viewing the structure of electrons in ethylene.

According to Lennard-Jones, the oxygen molecule (O=O) in its ground state consists of a double bond with four valence electrons. Of these there are two in the 2pσ state and two others in the {2pπ₊, 2pπ₋} state. As the first two 2pσ electrons are in the same state, their eigenfunction is symmetric within their coordinates and, according to the Pauli principle, antisymmetric for the spin. This means that they must have "opposing spin orientations." This consequently gives rise to a normal homopolar valence bond according to Heitler and London (spin valence). These two electrons form one of the valence lines of double bonding. The related charge distribution is rotationally symmetric around the O–O bond.

The two 2pπ electrons are quite different, eliciting another kind of valence bond from the first two 2pσ electrons. They form the other valence line of the double bond with an "opposite orbital moment" but a "parallel spin moment" (momentum valence).[51] Each occupies one of the degenerate $2p\pi_+$ and $2p\pi_-$ orbitals. The ground state of the oxygen molecule is accordingly a paramagnetic triplet state in conformance with experimentally observed paramagnetism of oxygen.

To arrive at ethylene from oxygen, Hückel took the second step in the approach sketched above, thereby obtaining formaldehyde ($O=CH_2$). The decomposition of the O nucleus into a C nucleus and two H nucleii and the newly formed C–H bonds modify the overall electron configuration of the double bond. Hückel took only qualitatively into account the related changes in energy of the corresponding σ and π electron states by perturbation, taking "known experimental facts" as a basis and refraining from calculating explicitly the corresponding integrals that he had previously defined. He regarded the σ and π electrons as independent of each other without taking their mutual interaction ("σ–π coupling") into closer consideration.[52] Hückel thus showed that, in the case of formaldehyde ($O=CH_2$) in the ground state, a normal homopolar valence bond (spin valence) is formed between the two σ electrons along with an analogous valence bond between the two π electrons. As a consequence of the perturbation from the two hydrogen substituents, there is a splitting of the doubly degenerate π one-electron state into two different

[50]Hückel, E.: *Zur Quantentheorie der Doppelbindung*, pp. 437–438.

[51]Cf. Heitler, W.: *Zur Quantentheorie der Valenz*, in: Naturwissenschaften, 17 (1929), 546–547.

[52]These and other drastic simplifications are now known as the "Hückel approximation." It was used by the Nobel laureates Ronald Hoffmann and Robert Woodward, among others, in the area of organic reactions and anorganic solids in systemizing the various findings. See Woodward, R. B., Hoffmann, R.: *Die Erhaltung der Orbitalsymmetrie*. VCH Verlagsgesellschaft, Weinheim, 1970.

states. They give rise to two polyelectronic states. One is symmetric in the coordinates of the electrons and hence according to the Pauli principle has to have an antisymmetric spin (oppositely directed moments of spin). It is a singlet state and diamagnetic. The other state is antisymmetric within the coordinates and therefore of symmetric spin (parallel moments of spin). It is a triplet and paramagnetic. Owing to the difficult calculations it was not yet possible to determine theoretically which of the two states was at a lower energetic level. So Hückel applied "experimental data" to distinguish the corresponding state. At the time no measurements were available on the magnetic susceptibility of formaldehyde, just of monosubstituted formaldehydes that were all diamagnetic. So Hückel reasoned by analogy from the "experimental data" that the O = C bond was diamagnetic in the ground state and was a singlet state. Hückel then demonstrated that this state in particular yielded the stability of the planar arrangement of formaldehyde and substituted formaldehydes. He also demonstrated that the energy of the π bond was smaller than the energy of the corresponding σ bond. Thus according to Hückel's interpretation, the O = C double bond differed from that of the O_2 molecule. This difference was due to the two hydrogen substituents bound to the carbon atom.

Finally, Hückel arrived at the last stage of his approach, progressing from formaldehyde to ethylene by separating two H nucleii from the second O nucleus. Once again the hydrogen substituents prefer a singlet ground state with a planar arrangement of the whole molecule. So the ground state is diamagnetically congruent with the experimental data of the day for ethylene and substituted ethylenes.[53] From there Hückel reached his goal: a quantum-mechanical interpretation of double bonding. Hückel listed two reasons for the rigidity of the double bond against rotation: "First, an energetic one: the σ electrons are more firmly bound. Second, a symmetry consideration: σ and π functions have different symmetries. The σ function is symmetric, the π function is antisymmetric with reference to the molecular plane."[54] The result of his quantum-mechanical model was thus that double bonds consist of two different types of bonds, one σ bond and one π bond. The "charged cloud" of σ electrons is distributed symmetrically around the C = C axis and has no stabilizing effect on the molecule's planar arrangement. On the basis of the above-described considerations Hückel showed that the inability of the substituents to rotate freely around the double bond originates from the asymmetric distribution of the π electrons around the axis of the C–C bond.

However, no completely convincing explanation for this could be found. In examining the hydrogen molecule Heitler and London[55] had eliminated the kinetic energy of the electrons by replacing it in Schrödinger's atomic equation with the difference between the atomic energy of H and potential energy terms. The result was an expression for the binding energy composed only of potential energy elements:

---

[53] Hückel, E.: *Zur Quantentheorie der Doppelbindung*, p. 453.

[54] *Chemiker im Gespräch: Erich Hückel*, p. 183.

[55] Heitler, W. und London, F.: *Wechselwirkung neutraler Atome und homöopolare Bindung nach der Quantenmechanik*, in: ZP 44 (1927), 455–472.

It obscured the fact that the kinetic energy contributes substantially to the covalent bond but is reduced by a conducive overlapping of simple orbitals. This was realized only a generation later by Klaus Ruedenberg.[56]

Hückel pointed out that the "charged clouds" of π electrons are at a maximum on a plane perpendicular to the plane of the "charged clouds" of the substituents and that a "rigidity" occurs between the plane on which the substituents' valences lie and the plane of the double bond. The character of the substituents does not affect this rigidity although its degree depends on them. But Hückel thought then that the reason for the rigidity lay in the mutual electrostatic repulsions of the "charged clouds" and could be explained classically. Nevertheless, even then he realized that the true cause of this charge distribution was a "quantum-mechanical effect" that could not be interpreted according to pure classical theory.[57] This also applied to the double bond between carbon and nitrogen (syn and anti isomerism for oximes), when the double-bonded atoms bear only three or two substituents, which are then forced onto the same plane.

Thus Hückel provided a physical explanation on the basis of quantum mechanics for the most important characteristic of van't Hoff's model of double bonding: the two perpendicular planes for the four substituents and the double bond, respectively. In his paper he wrote in this regard:

> Let the maximal planes of the charged clouds of the valence electrons binding the substituents be identified with the planes of the corresponding "valence lines," and let the charged clouds maximum of the double bond's $[\pi]_u$ electrons be identified with a preferred plane of the valence double bond, which lies perpendicular to the planes of the substituents. One sees that the picture provided by the theoretical interpretation of double bonding concurs broadly with the one van't Hoff had made for the rigidity of the double bond against rotation, in which he joined two rigid tetrahedrons along a common edge. However, the theory does not completely conform with this picture because no real meaning is attached to the four valence directions depicted by van't Hoff; only the plane on which they lie in van't Hoff's picture has such a meaning.[58]

Thus Hückel also deviated from van't Hoff's model of double bonding. He found out that the quantum-mechanical interpretation of the nonrotatability of the double bond led to a different understanding of the four valences (c, c', d, d') forming the double bond in van't Hoff's representation (Fig. 2.1). Hückel's result was that these four valences, which were the bent valences of double bonding in van't Hoff's visual model, were devoid of "real meaning." In a certain sense, Hückel's quantum-mechanical interpretation of double bonding was an early call for a critical verification against reality of the formulations in organic chemistry and the fundamental stereochemical postulates about tetrahedrally directed valence forces of the carbon atom. This recourse to quantum mechanics thus penetrated deep into the

---

[56] Ruedenberg, K.: *The Physical Nature of the Chemical Bond*, in: Review of Modern Physics 34 (1962), 326–376.

[57] Hückel, E.: *Zur Quantentheorie der Doppelbindung*, p. 455.

[58] Ibid. 454.

conceptual world and traditional visualizations of organic chemists. Hückel's criticism of van't Hoff's model was primarily aimed against the prevailing depiction of the carbon atom's directed valence forces. Put another way, quantum mechanics and its principles – specifically, applying electron spin, the Pauli principle and certain symmetry properties which characterize the eigenfunctions of the corresponding electron states of double bonding – led Hückel to a different understanding of double bonding. Moreover, it pointed to a new, less visual and more abstract understanding of stereochemical issues.

Finally, Hückel's quantum-mechanical interpretation of double bonding also afforded a qualitative explanation for the interconvertability of the cis trans isomers upon incidence of ultraviolet light or catalytic action. Such explanation meant consistent reorganization of these facts within the new conceptual framework. π electrons bind the atoms together in a double bond but more weakly than σ electrons. That is why a double bond is not twice as strong as a unique bond. π electrons are consequently more loosely bound than σ electrons.[59] When a π electron is excited by ultraviolet light, it loses its stabilization role in double bonding. Indeed, it has the opposite effect of working against the stabilization and becoming antibinding, allowing conversion to occur. Likewise, when a bromine atom is temporarily joined, it uses a π electron. The state of the second π electron is strongly perturbed, which can lead to a loss of molecular stability. It would have been desirable to be able to make quantitative predictions about the maximum energy levels at which the two stable cis and trans isomers emerge from each other. But, as Hückel stressed in his paper, it appeared hopeless to arrive at such information by purely theoretical means, given the methods of calculation available to him.

---

[59]In principle, electrons are indistinguishable. So it is conceptually incorrect to refer to σ and π electrons. Within the context of "models of independent electrons" (Hartree-Fock model, orbital model, model of the self-consistent field) the σ and π orbitals are what are meant. At the time no clean distinction was made between electrons and the orbitals constructed for their approximate description, and is still generally uncommon in the chemical literature even today. That is why philosophers have contended that the physicist's electron (i.e., electrons per se) and the chemist's electron (i.e., one-electron clouds or one-electron orbitals) were two different conceptual ideas. For more details, see Arabatzis, T.: *The Electron: A Biographical Sketch of a Theoretical Entity*, Princeton University, Ph. D. Thesis, 1995; *How the Electrons Spend their Leisure Time: Philosophical Reflections on a Controversy Between Chemists and Physicists*, in: *The Autonomy of Chemistry, 3rd Erlenmeyer – Colloquy for the Philosophy of Chemistry*, edited by Peter Janich and Nikolaos Psarros. Königshausen & Neumann, Würzburg, 1998, 149–159; *Representing Electrons: A Biographical Analysis of a Theoretical Entity*, The University of Chicago Press, Chicago, forthcoming; Arabatzis, T. and Gavroglu, K.: *The Chemists' Electron*, in: European Journal Physics 18 (1997), 150–163; Gavroglu, K.: *Appropriating the Atom at the End of the 19th Century: Chemists and Physicists at Each Other's Throat*, in: *Philosophers in the Laboratory*, edited by Valeria Mossini. Academia Nazionale di Scienze, Lettere e Arti di Modena, MUSIS, Roma 1996, 93–106; *The Physicists' Electron and Its Appropriation by the Chemists*, in: *Histories of the Electron, The Birth of Microphysics*, edited by Jed Z. Buchwald and Andrew Warwick. The MIT Press, Massachusetts, 2001, 363–400; Nye, M. J.: *Remodeling A Classic: The Electron in Organic Chemistry, 1900–1940*, in: Ebd., 339–361; Schwarz, W. H. E.: *Messung von Orbitalen und Schrödinger – Katzen*, in: Bunsen – Magazin 3 (2001), 59–68.

A year later, at the beginning of 1931, Linus Pauling and John Slater developed independently of each other a quantum-mechanical description of ethylene that completely confirmed van't Hoff's pictorial representation of double bonding.[60] In contrast to Hückel, Pauling visualized the two C = C bonds between the carbon atoms as two equivalent entities formed by an overlapping of two tetrahedrally directed sp^3 hybrid orbitals. Pauling underscored in his paper the quantum-mechanical legitimization of the carbon atom's tetrahedral structure and the relevant visual conception held by chemists regarding the formation of double bonds:

> This calculation provides the quantum mechanical justification of the chemist's tetrahedral carbon atom, present in diamond and all aliphatic carbon compounds,(...).
>
> A double bond behaves differently, however. (...) Hence the energy of the double bond will depend on the relative orientation of the two tetrahedral carbon atoms, and will be a maximum when the two sets of eigenfunctions show the maximum overlapping. This will occur when the two tetrahedral carbon atoms share an edge. Thus we derive the result, found long ago by chemists, that there are two stable states for a simple compound involving a double bond, a *cis* and a *trans* state, differing in orientation by 180°. *There is no free rotation about a double bond.*[61]

Pauling's quantum-mechanical interpretation of double bonding was grounded not only in his own idea of the "change in quantization of bond eigenfunctions" but also in the principle of maximum overlapping of the "bond eigenfunctions."[62] It is worth mentioning that Pauling's paper only cites Hückel's interpretation of double bonding in passing, in a footnote. On the basis of the concepts discussed above he expressed his conviction that his own interpretation of double bonding was more to the point than Hückel's:

> A discussion of rotation about a double bond on the basis of the quantum mechanics has been published by E. Hückel, which is, I feel, neither so straightforward nor so convincing as the above treatment, inasmuch as neither the phenomenon of concentration of the bond eigenfunctions nor that of change in quantization is taken into account.[63]

So Pauling's criticism of Hückel focused on his own concepts in explanation of the nonrotatability of double bonds. He offered no reason for why Hückel's explanation was less convincing than his. No doubt, Pauling could have phrased his criticism

---

[60]Pauling, L.: *The Nature of the Chemical Bond. Application of Results Obtained from the Quantum Mechanics and From a Theory of Paramagnetic Susceptibility to the Structure of Molecules*, in: JACS 53 (1931), 1367–1400; *The Nature of the Chemical Bond III: The Transition from One Extreme Bond Type to Another*, in: JACS 54 (1931), 988–1003; Slater, J. C.: *Directed Valence in polyatomic Molecules*, in: Physical Review 37 (1931), 481–489; *Molecular Energy Levels and Valence Bonds*, in: Physical Review 38 (1931), 1109–1114. Cf. Park, B. S.: *The Contexts of Simultaneous Discovery: Slater, Pauling, and the Origins of Hybridisation*, in: Studies in History and Philosophy of Science 31B (4) (2000), 451–474.

[61]Pauling, L.: *The Nature of the Chemical Bond*, 1378–1379. (Emphasis in original).

[62]Cf. Simoes, A.: *Converging Trajectories, Diverging Traditions: Chemical Bond, Valence, Quantum Mechanics and Chemistry*, 1927–1937, pp. 141–143.

[63]Pauling, L.: *The Nature of the Chemical Bond*, 1378–1379, note 25.

more skillfully. But this illustrates Pauling's tendency to assign a dominant role to his own ideas in the new discipline of quantum chemistry.[64]

Slater explained the formation of the double bond in the ethylene molecule in a way similar to Pauling, following van't Hoff. His advocacy of his own interpretation over Hückel's was certainly more diplomatically put: "This explanation seems much more convincing than the ingenious theory of Hückel, based wholly on *p* electron valence".[65]

There is a notable difference between Hückel's model of double bonding and the ones by Pauling and Slater. According to the latter, both C–C bonds are equivalent; not so with Hückel. Pauling and Slater justify the visual models of traditional chemists through skillful application of the mathematical language of quantum mechanics. Hückel's model, on the contrary, underpinned the organic chemist's classical visualizations while placing it at the same time on a new foundation that would ultimately point to new epistemological horizons.[66]

What Hückel had not taken into account in his quantum theory of double bonding was the nature of C–H bonds. Slater aptly stressed in his first paper: "The weakness of Hückel's theory is that he does not consider carefully the nature of the other bonds than the double one, and hence does not notice the fact which we point out, that the s valence must be combined with the p's to obtain a symmetrical arrangement."[67] Hückel never took up this issue himself, leaving it to others to investigate and elaborate.

Subsequently in a series of papers Robert S. Mulliken critically assessed and expanded Hückel's model of double bonding and Pauling's and Slater's concept of quantum-mechanical hybridization. Two years later Mulliken found a more satisfactory quantum theory of double bonding than Hückel, Pauling and Slater, using group theory.[68] Another pioneer in this area of research was W.C. Penney from England,

---

[64]He acted in a similar manner regarding Hückel's quantum mechanical treatment of the benzene issue. See here Section 3.2.

[65](Original English). Slater, J. C.: *Directed Valence*, p. 487.

[66]In the original, simple VB model elaborated by Heitler, London, Pauling and Slater, as well as Hückel, the atomic orbitals have to be suitably prepared in order to be able to construct the model of the molecular binding state. The suitable atomic orbital hybrids depended on the molecular situation and the success of the model being constructed. For doubly bonded systems, tetrahedral sp^3 hybrid orbitals, which produce two banana bonds, as well as sp^2 hybrid orbitals and $p\pi$ orbitals, which produce a σ bond and a π bond, are equally suitable. Historically, the first option was initially used by Pauling. The MO model developed by Hund, Mulliken and likewise by Hückel, who came to prefer this model, concedes complete equivalence between the different orbital representations. But it was only in the coming decades that it became quite clear that even in the MO model a double bond may be depicted either as two similar banana bonds or as a σ bond and a π bond. Not all chemists are aware of this even today. In Hückel's day only the latter σ/π representation was thought evident.

[67]Slater, J. C.: *Directed Valence*, p. 487.

[68]Mulliken, R. S.: *Electronic Structures of Polyatomic Molecules and Valence. I*, in: Physical Review 40 (1932), 55–62; *Electronic Structures of Polyatomic Molecules and Valence. II. General Considerations*, in: Physical Review 41 (1932), 49–71; *Electronic Structures of Polyatomic Molecules and Valence. III. Quantum Theory of the Double Bond*, in: Physical Review 41 (1932),

who developed the "right angle" model in 1934. He conducted a comprehensive comparison between his model and the other models already mentioned.[69]

2.1.5 The Response to Hückel's Quantum Theory of Double Bonding

In 1930 Hückel got the opportunity to present his quantum theory of double bonding in Königsberg at the annual meeting of the German Physical Society. But luck was not with him in this first attempt at publicizing his new research agenda before his professional colleagues. His talk was scheduled as the last of all the morning sessions, just before lunch, because – as Hückel remarks in his autobiography – his contribution was regarded as the "least important."[70] This was just one reason why he got no feedback. Much later Hückel still recalled that disastrous talk and his exasperation at the time:

> When my turn finally came at the beginning of the noon break, the ravenous participants got up one by one to go to lunch. Disorder, commotion and noise reigned in the auditorium. That annoyed and irritated me so much that I lost my bearings and the talk fell flat.[71]

A second chance came for Hückel at the XXXVth convention of the Deutsche Bunsen Gesellschaft from May 28 to June 1, 1930 in Heidelberg. The guiding theme of the conference was "Spectroscopy and molecular structure." Hückel once again introduced his quantum-mechanical treatment of double bonding, this time before the leading members of the nascent discipline of quantum chemistry and important German chemical physicists.[72] The title of his focus talk was: "On the quantum theory of double bonding and its stereochemical behavior."[73] Inspiring conversations with Hund and Heitler arose in Heidelberg as a result. Hückel also conversed with Mulliken, who presented another focus talk on 'Electronic states and chemical

---

751–758; *Quantum Theory of the Double Bond*, in: JACS 54 (1932), 4111–4112; *Electronic Structures of Polyatomic Molecules and Valence. IV. Electronic States, Quantum Theory of the Double Bond*, in: Physical Review 43 (1933), 279–302. Cf.: Park, B. S.: *Computations and Interpretations: The Growth of Quantum Chemistry, 1927–1967*, Chap. 3. (2) The Molecular Orbital Interpretation of Valence, 133-155; Simoes, A.: *Converging Trajectories, Diverging Traditions: Chemical Bond, Valence, Quantum Mechanics and Chemistry, 1927–1937*, Chap. 21. The Development of Mulliken's Program: What Are Electrons Really Doing in Molecules? 173–181.

[69] Cf. Berson, J. A.: *Erich Hückel – Ein Pionier der Organischen Quantenchemie: Leben, Wirken und späte Anerkennung*, Kap. 4.4. Die Hybridisierung in Doppelbindungen.

[70] Hückel, E.: *Ein Gelehrtenleben*, p. 134.

[71] Ibid.

[72] His audience included: K. F. Bohhoeffer, K. Clusius, P. Debye, J. Errera, A. Farkas, J. Frank, P. Harteck, F. Hund, W. Heitler, H. Mark, R. Mecke, R. S. Mulliken, K. Ramsauer, A. Smekal, G. Tammann, W. Weizel, L. Wolf.

[73] Hückel, E.: *Zur Quantentheorie der Doppelbindung und ihres stereochemischen Verhaltens*, in: Z. Elektrochem. 36 (1930), 641–645.

bonding in diatomic molecules.'[74] So this debut of Hückel in his new field was generally well received. There were intense discussions with the established quantum theorists, also concerning general issues like devising a consistent nomenclature for molecular states.[75] The expressions Hückel preferred to use in his papers for the pertinent states of π electrons were the following: $[\pi]_g$ and $[\pi]_u$. The indices g and u reflected the even [*gerade*] or odd [*ungerade*] character of the eigenfunction with reference to the molecule's center of symmetry.

Debye was delighted with Hückel's novel achievement. He referred to Hückel's "blossoming out" during a meeting in Leipzig with the director responsible for Europe, Tisdale, of the International Education Board.[76]

Lennard-Jones also wrote Hückel an enthusiastic letter about the paper from Bristol. He was primarily impressed with the way Hückel had derived the double bond between two carbon atoms from the known situation with the oxygen molecule:

> I was most grateful for your letter and your summary of your recent work. I was able to give an account of it in my lecture at Oxford last Monday night and everyone seemed very interested in it. Your method of deriving the carbon double bond from oxygen is most ingenious and provides a natural method of explaining the rigidity of the body and the formation of isomers. You will notice from my paper that the carbon molecule (C_2) formed from divalent carbon, should also have this peculiar type of linkage. Your work is a most valuable extension, which should be of great interest to chemists.[77]

In Germany the number of chemists keeping abreast of applications of quantum mechanics to their field was still very low. They showed particular interest in the explanation Hückel's paper offered for the stability of the spatially arranged substituents of C = C double bonding such as C = N, hence cis trans isomerism and syn and anti-isomerism. I mention in this regard, besides Hückel's brother Walter, the chemists Hermann (later Herman) Mark and Friedrich Ebel.

2.1.5.1 Hermann Mark's Interest in Hückel's Quantum Theory of Double Bonding

Mark was born on May 3, 1895 in Vienna to a Jewish medical doctor. After pursuing studies in chemistry and receiving his doctorate in 1921 in his home town, he moved to Berlin to begin work as assistant in chemistry at the local university. 1922 he was a fellow researcher at the Kaiser Wilhelm Institute of Fiber Research in Berlin, where he was promoted to department head three years later. This institute was the first to conduct X-ray investigations of polymers and determine their structures. 1927

---

[74] Mulliken, R. S.: *Elektronenzustände und chemische Bindung in zweiatomigen Molekülen*, in: Z. Elektrochem. 36 (1930), 603–605.

[75] Friedrich Hund, conversation with A. Karachalios. Göttingen August 22, 1994.

[76] According to Tisdale, Debye said: "Hückel is blossoming out." RAC, Box 52, Folder 805, P. 39, from Doktor Tisdale's log., January 22, 1930, Leipzig.

[77] SBPK, Paper of Erich Hückel, Box 6, Folder 5.111, J. E. Lennard-Jones an Hückel, 29. January, 1930.

Mark was engaged as a researcher to work under Kurt H. Meyer in the main laboratory of the dye syndicate I.G. Farben in Ludwigshaven. He subsequently headed the department of polymer chemistry there.[78] At Ludwigshafen he applied techniques taken from physics to polymer research, notably X-ray and electron diffraction. His research team comprising organic and physical chemists examined the structure of polymers like polystyrene and polyvinyl chloride and worked on their synthesis.[79]

Meyer had left Mark the freedom also to work on other research not directly related to industrial applications. So Mark was able to conduct the first electron-diffraction analyses on pure gases at the I.G. Farben laboratory. He determined the binding distances and angles of a whole range of aliphatic and aromatic compounds. The free rotatability of simple carbon bonds was another problem that occupied him. Pauling visited Mark in Ludwigshafen in 1930 and was very impressed with his accomplishments in physical chemistry.[80] Another enthusiastic admirer was the young scientist Teller, who also visited Mark in the main laboratory of I.G. Farben. Teller was fascinated by Mark's expertise extending beyond organic and physical chemistry into quantum mechanics. Teller later told about Mark's scientific ideas: "It was all so new, and yet it was in the middle of chemistry and mathematics." Besides that, Teller remarked, Mark was also "a very wonderful man."[81] His knowledge of quantum mechanics interested Teller to the point that he abandoned chemistry in favor of studying physics.[82]

Mark's own interest in quantum theory as a tool for solving problems in chemistry arose from contacts he had with pioneers of the newly formed discipline of quantum chemistry. His friendship with Mulliken began in 1925 when they both joined a hiking tour in the Dolomites with mutual friends.[83] In 1930 Mark also attended the Bunsen Society's XXXVth convention mentioned earlier and presented a talk on "Determining molecular structure by electron diffraction in gases."[84]

---

[78]Herman Francis Mark, in: *Lexikon bedeutender Chemiker*, Verlag Harri Deutsch, Thun; Frankfurt am Main, 1989.

[79]Reinhardt, C.: *Basic Research in Industry: Two Case Studies at I. G. Farbenindustrie AG in the 1920s and 1930s*, in: Travis A. S., Schröter H. G., Homburg, E.: *Determinants in the Evolution of the European Chemical Industry, 1900–1939: New Technologies, Political Frameworks, Markets and Companies*. Kluwer, Dordrecht 1999, 74–76; Furukawa, Y.: *Polymer Science: From Organic Chemistry to an Interdisciplinary Science*, in: *Chemical Sciences in the 20th Century, Bridging Boundaries*, edited by Carsten Reinhardt. Wiley-VCH, Weinheim, 2001.

[80]Morawetz, H.: *Herman Mark, Life and Accomplishments*, in: Macromol. Symp. 98 (1995), 1173–1184.

[81]Cited from Blumberg, S. A. and Owens, G.: in: *Energy and Conflict, The Life and Times of Edward Teller*. G. P. Putnamś Sons, New York, 1976, 34.

[82]Ibid.

[83]Mulliken, R. S.: *Life of a Scientist*. Springer-Verlag, Berlin, Heidelberg, New York, London, Paris, Tokyo, 1989, pp. 54 and 74.

[84]Mark, H.: *Die Ermittlung von Molekülstrukturen durch Beugung von Elektronen an einen Dampfstrahl*, in: Z. Elektrochem. 36 (1930), 675–676.

In 1932 Mark followed the advice to leave I.G. Farben[85] and that fall accepted a call as full professor of physical chemistry at the University of Vienna as successor to Rudolf Wegscheider. At Vienna he founded the first institute for interdisciplinary research and teaching in polymer chemistry in the Austrogerman region.[86]

In the Austrian capital Mark continued to pursue his interest in quantum mechanics. He published a paper in the following year on "The quantum-mechanical interpretation of valences." It appeared in the comprehensive handbook "Stereochemistry, a summary of the results, foundations and problems."[87] This handbook was edited by a pupil of Emil Fischer, Karl Freudenberg, who occupied the chair for chemistry at the University of Heidelberg and directed the chemistry laboratory.[88] As editor, Freudenberg attempted to draw under a single umbrella reports by leading researchers at the forefront of all branches of chemistry. He opened his foreword with an emphasis on the diversity of stereochemical findings to date: "When Victor Meyer introduced the term stereochemistry, the variety of paths along which this field of knowledge would expand and affect fields ranging from serology to wave mechanics were hidden even from his far-sighted purview."[89]

In his report Mark introduced the basic concepts of quantum theory in a language easily understood by chemists in order to describe stereochemistry in quantum-mechanical terms. He outlined the tetrahedral structure of the C atom and the rigidity of double bonds, dwelling on qualitative aspects and referring to the article by his fellow chemist Friedrich Ebel for a more detailed treatment.

2.1.5.2 Friedrich Ebel's Dissemination of Hückel's Quantum Theory of Double Bonding

Ebel was born on February 3, 1901 in Reichelsheim, Odenwald. After completing his primary and secondary education in Kaiserslautern he began his courses of study in chemistry in 1920 at the University of Munich under Richard Willstätter. In January 1924 he started writing his dissertation under the supervision of the private

[85] The historian of science Ute Deichmann gave the following reason in her detailed and well-documented study on chemists and biochemist in the Nazi period: "Herman Mark left the I.G. Farben in 1932 after he had received the advice from the chairman of the board Dr. W. K. Friedrich Gaus, to keep an eye out for a university position outside of Germany. If Hitler rose to power, Gaus said, Mark would not be promoted." See Deichmann, U.: *Flüchten, Mitmachen, Vergessen: Chemiker und Biochemiker in der NS-Zeit*. Wiley-VCH, Weinheim, 2001, p. 183.

[86] A few years later Mark joined many other scientists in exile for political reasons during the period of National Socialism. He first stayed in Switzerland then emigrated via England to Canada and finally to the USA. On Mark's scientific career and his fate see Deichmann, U.: *Flüchten, Mitmachen, Vergessen*, Chap. 4.2.2.

[87] *Stereochemie, eine Zusammenfassung der Ergebnisse, Grundlagen und Probleme*, Herausgegeben von K. Freudenberg. Franz Deuticke, Leipzig und Wien 1933.

[88] On K. Freudenberg, see Meinel, C.: *Freudenberg, Karl Johann*, in: *Dictionary of Scientific Biography,* Supplement II, ed. von Frederick L. Holmes, Ch. Scribner's Sons, New York, 1990.

[89] *Stereochemie*, ed. K. Freudenberg, Vorwort.

lecturer Richard Kuhn (who received the Nobel Prize in 1938)[90] and was conferred his doctorate in the following year *summa cum laude*. Afterwards he was Kuhn's assistant, who had been appointed professor of general and analytic chemistry at the Federal Polytechnic in Zurich in the fall of 1926. At Zurich Ebel started doing independent research and received his permission to teach in the summer of 1928 on the basis of his habilitation thesis "On the foundations of adhesion theory."[91] The topics of his lectures in the summer semester of 1928 and the following winter term were "Analytic and synthetic methods of organic chemistry" and "The mechanism of chemical reactions." For the summer semester in 1929 he received a teaching assignment on "Coal hydrogenation and other contact catalytic processes of organic chemistry."[92]

At that time his teacher in Zurich, Kuhn, was working on the stereochemistry of ortho-substituted biphenyls. He realized that the free rotation of the substituents was impeded by space constraints and introduced the term "atrop-isomerism."[93] In 1929, the same year that Kuhn was appointed director of the institute of chemistry at the Kaiser Wilhelm Institute of Medical Research in Heidelberg, Ebel started working as of October in the main laboratory of the chemical company BASF, where he remained up to retirement.

Ebel probably took an interest in stereochemistry during his period at Zurich under Ebel's influence. He was specifically attracted to the problems posed by carbon compounds. In book two of the handbook on stereochemistry[94] mentioned above, two longer reports by Ebel appeared under the titles "The theory of tetrahedrons" and "Phenomena of isomerism."[95] Ebel's contributions attempted to draw the large amount of data on isomeric phenomena and the tetrahedral theory by van't Hoff and Le Bel into closer association with the latest advances in the areas of physical chemistry and quantum mechanics.

In the first contribution Ebel demonstrated that physical methods like interference analysis, dipole measurements, light scattering, absorption and Raman spectroscopy confirmed the classical tetrahedral theory. In his second contribution, however, he discussed some "polymorphic phenomena," for instance, cinnamic acid and chalco-substances, in which the limitations of van't Hoff's model of double bonding were first clearly ascertainable. These and other deficiencies in the existing models served

---

[90]R. Kuhn's further political and scientific destinies are discussed in Deichmann, U.: *Flüchten, Mitmachen, Vergessen*.

[91]UA-BASF, W1 (Personenarchiv) Friedrich Ebel.

[92]Ibid.

[93]There is a detailed description of R. Kuhn's concept of "atrop-isomers" or "atrop-isomerism" in Freudenberg's *Handbuch der Stereoisomerie*, 2. Buch, Kap. VI., pp. 803–822. Kuhn authored the connected essay on "Molecular asymmetry."

[94]The entire handbook was divided into three books. The first carried the subtitle "General and physical stereochemistry," the second, "Stereochemistry of carbon compounds" and the third "stereochemistry of the elements other than carbon."

[95]Ebel, F.: *Die Tetraedertheorie*, in: *Stereochemie*, pp. 525–551; *Die Isomerieerscheinungen*, in: Ibid., pp. 553–661.

as the starting point for Ebel's discussion of Hückel's quantum-theoretical model of double bonding, which he outlined very clearly in his second contribution.

The outstanding point in Ebel's discussion of the visual depiction of the ethylene double bond was the novel nature of Hückel's theory against van't Hoff's model. He explicitly stated that in Hückel's quantum-theoretical model the four valence lines forming van't Hoff's double bond had "no real meaning." He continued: "If a valence line must be drawn, then at most, let it be one corresponding to the spin valence."[96] Hückel's quantum theory of double bonding thus gained the recognition of chemists and made its first inroads toward acceptance in chemistry.[97]

2.1.6 Lecturer of "Chemical Physics" in Stuttgart: 1930–1937

At the end of September 1930 Hückel's fellowship with the Notgemeinschaft also expired, leaving him in a difficult situation. He had started to reconnoitre his new field of research located between physics and chemistry yet beyond the bounds of traditional physical chemistry. The new marginal area of quantum chemistry was still in its formative stage in Germany. As Hückel aptly pointed out in his autobiography, there were "still no professorships and accordingly no lecturers and assistants" in this nascent discipline.[98] He described his unpleasant predicament as follows: "So there I stood between two stools, physics and chemistry, both of which were out of the question for me, and between them: a gaping void."[99]

Debye had great expectations of Hückel's new line of research and once again came to the rescue.[100] With the support of a few professors at the Technische Hochschule in Stuttgart, such as the theoretical physicist Peter Paul Ewald, the physical chemist Georg Grube and the experimental physicist Paul Regener, he arranged for a teaching assignment for Hückel as *Dozent* of "chemical physics" starting on October 1, 1930. Hückel reports about this lectureship in his autobiography:

> There was Physics; there was Chemistry. There was also Physical Chemistry. But what I was doing did not fit into any of these slots. Even Physical Chemistry as it was being conducted in those days was something quite different from what I was doing. My occupation was applying a new theory in physics, "Quantum Ohysics," to problems in chemistry. To distinguish between this new border area from conventional Physical Chemistry, one used (and

---

[96]Ibid.

[97]Stefan Goldschmidt (1889–1971), full professor of organic chemistry at the Polytechnic in Karlsruhe from 1927 also mentioned Hückel's theoretical treatment of double bonding in his comprehensive handbook on stereochemistry. Cf. *Hand- und Jahrbuch der chemischen Physik*, edited by von A. Eucken und K. L. Wolf, Band 4 *Stereochemie* von Stefan Goldschmidt. Akademische Verlagsgesellschaft M. B. H., Leipzig, 1933, p. 73.

[98]Hückel, E.: *Ein Gelehrtenleben*, p. 137.

[99]Ibid.

[100]Tisdale wrote in his notebook under the date 06.02.1931: Hückel; is now in Stuttgart with Ewald. Debye thinks that H. will fulfill all the promises which H. has evidenced up to now, in RAC, Box 52, Folder 805, from Doctor W. Tisdale's log. – Leipzig. February 6, 1981.

still continues to use) the expression "Chemical Physics," an expression I do not consider particularly well chosen because it actually implies the application of chemical methods to physics and this is not the case. That is probably why the term "Quantum Chemistry" came into more common use. It expresses much better the application of methods in physics – to be specific, quantum theory – to chemical problems. This border area of science was at the beginnings of its development.[101]

Hückel was not happy with this new teaching assignment. His position at the polytechnic in Stuttgart was not permanent. It was financed – and badly at that – by a few grants. At the end of each semester it was never certain that the succeeding term would be approved. There was a limiting clause in his teaching contract that stipulated that he only receive his full allowance if no less than three auditors sign up for his lectures. Hückel later described these degrading working conditions at Stuttgart as well as the solidarity he encountered among his students: "It happened that individual students – like the son of Professor Regener – registered themselves officially in my course just so that I would not lose a part of my livelihood."[102] The low attendance at Hückel's lectures must have quite certainly been a constant source of worry for him. It boded ill for the continuation of his teaching assignment.

Looking back on this period as lecturer in Stuttgart Hückel recalled: "I remained in these desperate circumstances in Stuttgart, which were psychologically very taxing for me, from 1930 to 1937. Anne spoke, and still speaks of our 'seven years of shame.' Later it would turn out that they were the most productive years of my life in science."[103]

During these most productive years of his scientific career in Stuttgart, Hückel offered a joint discussion seminar with Paul Ewald, a pupil and friend of Sommerfeld. By that time Ewald was already an international authority in the new area of crystalline structure analysis. Their course on atomic physics was offered every academic year between 1931 and 1935.[104] In the same period Hückel offered a number of lectures, some of them even including practice sessions, in which he concentrated on the relations between quantum theory and chemical valence or molecular structure.[105] In the lecture "Molecular structure and physical properties," for example, which he taught during the academic year 1935/1936, Hückel discussed the physical properties of the chemical substances in relation to their chemical structure in view of recent findings in quantum mechanics on valence and

[101] Hückel, E.: *Ein Gelehrtenleben*, pp. 136–137.

[102] Ibid.

[103] Ibid.

[104] Cf. *Programm der Wüttembergischen Technischen Hochschule Stuttgart*, Studienjahr 1931/1932, 1932/1933, 1933/1934, 1934/1935, Greiner & Pfeifer, Druckerei u. Verlagsanstalt, Stuttgart.

[105] The titles of his lectures were: Introduction to the theory of spectra II: molecular spectra (1931/1932); Wave mechanics I (1932/1933); Physical properties and chemical constitution (1932/1933); Problems of reaction rate (1932/1933); The periodic system in light of atomic theory (1933/1934); Physical properties and chemical constitution (1934/1935); Quantum theory of molecular forces, specifically the theory of valence (1934/1935). Taken from the course catalogs: Programm der Wüttembergischen Technischen Hochschule Stuttgart.

the electronic structure of diatomic molecules, i.e., Heiter's and London's scheme of spin valence.[106]

While Hückel was at Stuttgart, the first edition of Walter Hückel's two-volume work "Theoretical foundations of organic chemistry" was published in 1931.[107] It was partly a joint effort of the two brothers. Its interdisciplinary flavor spoke as well to chemists as it did to physicists. Erich wrote the section on quantum mechanics and assisted in preparing the sections on physics and organic chemistry.[108] This work applied to chemistry theories normally associated with physics, including quantum mechanics. Walter Hückel emphasized in the foreword: "The second volume then conveys information about the theoretical tools provided by recent developments in physics and physical chemistry for deeper insight into problems that research in pure chemistry has not succeeded in solving."[109] With this work the Hückel brothers evidently intended to bolster physical chemistry in the area of organic chemistry. They outlined the approaches prevalent in both fields and attempted to breach the existing gap between them. For German-speaking regions Walter's work was a path-breaking handbook, less for what later came to be known as "organic quantum chemistry" than for the field of "physical organic chemistry."[110] The way to "organic chemistry" was paved principally by Erich's own papers, written during his "seven years of shame" in Stuttgart.[111]

---

[106]SBPK, Paper of Erich Hückel, Box 4, Folder 1.122, Vorlesungsmanuskript: Molekularstruktur und physikalischen Eigenschaften, Fragment, 69 Blätter, 1935/1936.

[107]Hückel, W.: *Theoretische Grundlagen der Organischen Chemie*. Erste Auflage, Band I, II, Akademische Verlagsgesellschaft M. B. H., Leipzig, 1931.

[108]Walter Hückel notes in the foreword to the work: "I must express my hearty thanks above all to my brother Erich Hückel for his help and collaboration. He not only authored the section on quantum mechanics but also served as my constant advisor and aid in writing the chapters on physics, also giving me many valuable tips for the sections on organic chemistry." Hückel, W.: *Theoretische Grundlagen der Organischen Chemie*, Band I, 1931, Vorwort, p. VIII.

[109]Ibid., p. III

[110]On the rise and development of "physical organic chemistry," see Cerruti, L.: *Free Electrons, Lo sviluppo della chimica organica fisica, 1900–1940*, in: F. Abbri e M. Ciardi (a cura di), Atti del VIII Convegno Nazionale di Storia e Fondamenti della Chimica (Arezzo, 28–30 Ottobre 1999), Roma, Accedemia Nazionale delle Scienze detta die XL, 1999, 207-263; Gortler, L.: *The Physical Organic Community in the United States, 1925–50. An Emerging Network*, in: Journal of Chemical Education 62 (1985), 753–757; Morris, P. J. T., Travis, A. S. and Reinhardt, C.: *Research Fields and Boundaries in Twentieth-Century Organic Chemistry*, in: *Chemical Sciences in the 20th Century. Bridging Boundaries*, edited by Carsten Reinhardt. Wiley-VCH, Weinheim, 2001; Nye, M. J.: *From Chemical Philosophy to Theoretical Chemistry. Dynamics of Matter and Dynamics of Disciplines, 1800–1950*. University of California Press, Berkeley, 1993, Kap. 7–9; Saltzman, M.: *James Bryant Conant and the Development of Physical Organic Chemistry*, in: Journal of Chemical Education 49 (1972), 411–412; Saltzman, M.: *The Development of Physical Organic Chemistry in the United States and the United Kingdom: 1919–1939, Parallels and Contrasts*, in: Chemistry in Britain 63 (1986), 588–593; Stocklöv, J.: *Arthur Hantzsch: Wegbereiter der physikalischen organischen Chemie*. Dissertation, Halle (Saale), 1996.

[111]With reference to the neologism "organic quantum chemistry" I rely on the paper by J. A. Berson: *Erich Hückel – Pionier der Organischen Quantenchemie: Leben, Wirken und späte Anerkennung*.

Erich Hückel's papers from Stuttgart treated the benzene problem and other unsaturated and aromatic compounds.[112] Mulliken later referred to the first of these papers, entitled "Quantum-theoretical contributions to the benzene problem. I. The electron configuration of benzene and related compounds"[113] as "monumental".[114] It served as his habilitation thesis at the polytechnic in Stuttgart in 1931 in his bid for permission to teach theoretical physics there. The faculty regulations stipulated the submission of a separate thesis before he could make such a disciplinary transition.[115]

Altogether, we can regard Hückel's lectureship at Stuttgart from October 1, 1930 to November 1, 1937[116] as the first attempt at a German university to implement quantum chemistry in research and teaching.

2.2 Quantum Theory of Aromaticity: Symmetry and Electronic Shell Completion

With his article on the quantum theory of double bonding finished and sent out to the editors of the *Zeitschrift für Physik* (it was received from Leipzig on January 9, 1930), Hückel moved on to a new research agenda that he would continue to pursue after his move to Stuttgart that October. He decided to tackle the unsolved question of the binding state of alternating unique and double bonds. This electron configuration characterized the carbon atoms in benzene and other aromatic compounds.

Walter Hückel had first pointed out this benzene problem to his brother, also giving him some useful tips and advice about certain properties of the benzene ring, which contemporary theory had not yet managed to explain fully.[117] Erich Hückel later recounted that everyone had cautioned him against working on the benzene problem as far too hard a nut to crack. Nonetheless Hückel was convinced that the problem "really ought to be solvable somehow."[118] As will be shown in the following chapters, Hückel subsequently not only delivered the solution to the benzene problem but also provided a novel impetus for a changed theoretical approach from classical organic chemistry.

---

[112]See Section 2.2.2.

[113]Hückel, E.: Quantentheoretische Beiträge zum Benzolproblem. I. Die Elektronenkonfiguration des Benzols und verwandter Verbindungen, in: ZP 70 (1931), 204–286.

[114]Mulliken, R. S.: *Molecular Scientists and Molecular Science: Some Reminiscences*, in: JCP 43 (1965), pp. 2–11, p. 8.

[115]According to Decree No. 9595 of the Württemberg Ministry of Culture, Hückel was officially habilitated on August 5, 1931. Personalbogen für Hückel Erich in: HSB, Bestand 310, PA Nr. 6230 (Erich Hückel).

[116]Ibid.

[117]*Chemiker im Gespräch: Erich Hückel*, p. 184.This conclusion also follows from the brothers' collaboration on Walter's book mentioned above, *Theoretische Grundlagen der Organischen Chemie*.

[118]Ibid.

2.2.1 Prehistory and Hückel's Point of Departure

Much space would be needed for a comprehensive history of the benzene problem and the concept of aromaticity from its origins to 1930. It certainly eludes summary description within the bounds of a few pages.[119] The following will therefore address Erich Hückel's approach to the benzene problem and aromaticity. The focus will be on chemical and physical aspects of particular relevance to Hückel's quantum-mechanical treatment of benzene and related substances with aromatic properties. Thus a better sketch is possible of Hückel's quantum-chemical profile in the contested no man's land between chemistry, physical chemistry and physics. It will also better reflect his self-designation as a quantum chemist.

In the introduction to his first paper on the benzene problem,[120] Hückel's guiding assumption is that the stereochemical arrangement of the benzene molecule in space is as good as settled. Kekulé's hypothesis of a regular hexagonal arrangement of the six carbon atoms on a plane concurred with observations of the chemical behavior of benzene. Kekulé's hypothesis was also confirmed by röntgenographic determinations of its crystalline structure and by an analysis of the atomic arrangement of benzene vapor exposed to electron waves.

But Hückel underscored that the spatial configuration of the atom was just one aspect of the benzene problem as a whole. The particular reactivity of benzene or other aromatic compounds remained puzzling when compared to other saturated and unsaturated carbon compounds. Kekulé's formulation of the binding state of carbon atoms in the benzene ring, based on the postulate of quadrivalence for carbon, could not explain why benzene and other aromatic substances permit substitution reactions much more readily than other unsaturated carbon compounds and undergo addition only under drastic conditions.[121] In the case of cycloalkene (cyclo-octatetraene), a

---

[119] A number of detailed descriptions of the benzene problem already exist in the history of science. The first part of the informative article by Stephen G. Brush on *Dynamics of Theory Change in Chemistry*, from 1999, provides an overview of the historical developments of the benzene problem as well as a critical discussion of the important stages along with further references. He mentions Hückel briefly as the researcher whose MO method contributed much to modern theoretical views on the benzene problem and other aromatic compounds without, however, discussing Hückel's approach. Cf. Brush, S. G.: *Dynamics of Theory Change in Chemistry: Part 1. The Benzene Problem 1865–1945*, in: Stud. Hist. Phil. Sci. 30 (1999), 21–79. As concerns the historical development of the concept of aromaticity, see footnote 121 below.

[120] Hückel, E.: Quantentheoretische Beiträge I. Die Elektronenkonfiguration des Benzols.

[121] On the origin and further evolution of the concept of "aromatic compounds" or "aromaticity," see Garratt, P. und Vollhardt, P.: *Aromatizität.* Thieme-Verlag, Stuttgart, 1973; Hauptmann, S.: *Aromatizität–Bedeutungswandel eines Begriffes*, in: Zeitschrift für Chemie 13 (1973), 361–364; Maier, G.: *"Aromatisch"-was heißt das eigentlich?*, in: Chemie in unser Zeit 9 (1975), 131–141; Schütt, H. W.: *Der Wandel des Begriffs "aromatisch" in der Chemie*, in: *Begriffswandel und Erkenntnisfortschritt in den Erfahrungswissenschaften. Kolloquium an der Technischen Universität Berlin, WS 84/85.* Universitätsbibliothek der Technischen Universität Berlin, Berlin 1985, 255–272; Schleyer, P. v. R., Jiao, H.: *What is Aromaticity?* in: Pure and Applied Chemistry 68 (1996), 209–218. Von neueren Arbeiten ist die folgende Diplomarbeit zu erwähnen: Neus, J.: *Aromatizität Geschichte und mathematische Analyse eines fundamentalen chemischen Begriffs.*

compound synthesized by Willstätter,[122] Kekulé's formulation of the bonding state with alternating unique and double bonds could not explain the absence of aromatic properties. Its behavior rather resembled unsaturated compounds.[123]

Hückel discussed such paradigmatic cases in the introduction to his paper. He realized that the so familiar "aromatic" character of benzene and its related compounds could not be reflected in Kekulé's views on molecular structure.[124] Some decisive element about the behavior of aromatic compounds was missing. So Hückel first started looking for a concept that would allow a distinction to be made between aromatic and non-aromatic behaviors of carbo- and heterocyclic compounds. Hückel explicated:

> It appears initially as if the number 6 for the C atoms were of significance for the unusual behavior of the benzene ring. Bamberger was the first to point out that this special number 6 plays a significant role, in the general sense, not just in benzene (and other carbocyclic compounds) but also in heterocyclic ring-shaped compounds (where the number of atoms in the ring need not be 6).[125]

Eugen Bamberger (1857–1932) began to study the special properties of aromatic compounds in the early 1890s. He used the centrally arranged formulas for benzene proposed by A. Baeyer, L. Meyer and H. E. Armstrong to explain the anomalous chemical behavior.[126] The six carbon atoms are arranged in a regular hexagon of simple bonds with six attached hydrogen atoms. The remaining six valences are directed toward the center of the ring, resulting in a completely symmetrical distribution of affinity.

Bamberger held the view that the six valences of affinity ("sextet of affinity") were responsible for the special properties of benzene. He generally thought that the key to the aromatic properties would be found in certain characteristics requiring the six valences of affinity. He called such systems "hexacentric." They included: "All hexacentric systems – whether they have five or six members, whether they be constructed of carbonic material alone or with other elements – are

---

Hyle Studies in History and Philosophy of Chemistry 2, Hyle Publications, Karlsruhe 2002. Internet: www.hyle.org/publications/books/neus; *Aromaticity*, in: Chemical Reviews, 101 (5) (2001), 1119–1150.

[122] Willstätter, R., Waser, E.: *Über Cyclo-octatetraen*, in: B. 44 (1911), 3423–3445; Willstätter, R., Heidelberger, M.: *Zur Kenntnis des Cyclo-octatetraens*, in: B. 46 (1913), 517–527.

[123] Cyclobutadiene is a similar case, eluding successful synthesis over many decades. This was as intensely pursued as in the case of cycloalkene, because according to Kekulé's conceptions and Pauling's VB model, one expected to find as pronounced a stability for benzene.

[124] Hückel, E.: Quantentheoretische Beiträge I. Die Elektronenkonfiguration des Benzols., pp. 206–207.

[125] Ibid., p. 207

[126] Bamberger, E.: *Beiträge zur Theorie sechsgliedriger Ringsysteme*, in: Liebig's Annalen der Chemie 257 (1890), 1–55; *Ueber die Constitution fünfgliedriger Ringsysteme*, in: B. 24 (1891), 1758–1764; *Zur Frage der Valenz des Pyrrolstickstoffatoms*, in: B. 26 (1893), 1946–1947; *Zur Constitution fünfgliedriger Ringsysteme*, in: Liebig's Annalen der Chemie 273 (1893), 373–379.

characterized by a specific 'aromatic type'[127] that definitely distinguishes them from (aliphatic and alicyclic) compounds lacking such potential bonds."[128] Hückel emphasized Bamberger's inclusion of six-membered and five-membered rings in such heterocyclic systems as pyridine or pyrrole, furane and thiophene.

Hence Bamberger's theory attempted to attribute to all compounds of "aromatic type" a system of six "potential" valences.[129] Using the language of classical organic chemistry, Bamberger described what would later be known as aromatic delocalized electron sextets, a concept developed by James Wilson Armit and Robert Robinson in 1925.

By applying the new electron theory of valence by Gilbert Newton Lewis to aromatic compounds, Armit and Robinson drew a connection between aromaticity and electron theory.[130] They ascribed the typical properties of aromatic compounds to the presence of a group of six electrons.[131] But they could not explain why the number six in particular was of such significance. The number six remained an unexplained axiom. It was Hückel who provided a satisfactory quantum-mechanical explanation. He closed the introduction to his paper with the following outline:

> The present paper will attempt a quantum-mechanical treatment of the special binding state of benzene as compared to the binding states of other nonsaturated ring systems and thus also to attain some *theoretical understanding of the aromatic character of cyclic systems* containing 6 electrons not assignable to simple pairs of bonds. It will be shown that, in a certain sense, *the number 6 of such electrons forming a ring* form a "*complete electron shell,*" while the numbers 4 and 8 do not deliver such a complete shell.
>
> The result, on its own, would not be completely satisfactory, however, if this *quantum theoretical "model"* were not in a position to indicated the properties of the benzene ring known hitherto only from observations of the behavior of benzene.[132] [...]
>
> I only mention here that, on the basis of electron configurations of benzene treated in the present paper, *a quantum-theoretical explanation of the chemical data* is possible *by heuristic means*, and that furthermore it appears as if *a quantum-theoretical approach can also eliminate certain problems and incongruencies in current conceptions.*[133]

---

[127] Jean-Baptiste Dumas introduced the concept of chemical type in 1840. See Dumas, J. B.: *Ueber die Constitution der Essigsäure und Chloressigsäure*, in: Annalen der Chemie und Pharmacie 33 (1840) 179–182; *Ueber das Gesetz der Substitutionen und die Theorie der Typen*, in: Annalen der Chemie und Pharmacie 33 (1840), 259–300. Vgl. dazu auch Brock, W. H.: *The Norton History of Chemistry*. W. W. Norton & Company, New York 1992, Chap. 6 Chemical Method, 210–240.

[128] Cf. Bamberger, E.: *Ueber die Constitution fünfgliedriger Ringsysteme* (1891), p. 1762.

[129] Cf. Bamberger, E.: *Beiträge zur Theorie sechsgliedriger Ringsysteme* (1890), p. 47.

[130] Armit, J. W., Robinson, R.: *Polynuclear Heterocyclic Aromatic Types. Part II. Some Anhydronium Bases*, in: Journal of the Chemical Society (London) 127 (1925), 1604–1618.

[131] " (. . .) the posession of such [electron] groups, confers chemical stability as shown, for example, by reduced unsaturation and a tendancy to maintain the type. These are, of course, the chief characteristics of benzenoid systems, and here the explanation is obviously that six electrons are able to form a group which resists disruption, and may be called the *aromatic sextet*." (Emphasis in the original English), Ibid., p. 1604 f.

[132] Hückel meant here the chemical behavior of benzene upon multiple substitution and the applicable chemical rules.

[133] Hückel, E.: Quantentheoretische Beiträge I. Die Elektronenkonfiguration des Benzols, p. 213. (My emphasis).

The emphasis Hückel placed on the "aromatic character of the cyclic systems" is striking. He was basically following the approach of an organic chemist, who is primarily interested in the relations within classes of substances, in this case aromatic substances, and their consequences.[134] Correspondingly, Hückel regarded benzene as an archetype at the top of the class of aromatic compounds. He was evidently looking for a criterion for aromaticity not just as a distinguishing factor between aromatic and non-aromatic compounds but also as a means to classify a compound as aromatic. The number of electrons are conspicuously in Hückel's explicatory focus as opposed to any structural properties.[135] Finally, the above quote indicates Hückel's clear intention to propose a "quantum-theoretical model" of benzene. Indeed, his purpose reached beyond theoretical explanation. His quantum-mechanical considerations were supposed to enhance the empirical knowledge in chemistry about aromatic compounds.

2.2.2 Quantum-Mechanical Interpretation of the Electronic Configuration of Benzene and Related Aromatic Compounds

2.2.2.1 Hückel's Two Approximative Statements

The underlying idea of Hückel's quantum-mechanical treatment of benzene and its aromatic derivatives was that the molecule and its electronic structure should be considered as an integral whole. An assessment based only its interactions with neighboring atoms was not enough. The guiding influence of the substituents in the benzene ring Hückel saw as suggestive of this approach.[136] It was already known from experiments on the chemical behavior of benzene that a substituent at one location of the ring directed a second substituent located elsewhere on the ring into an ortho, para or meta position, depending on the nature of the second substituent. This revealed to Hückel that a disturbance in the distribution of charges among the electrons at one place on the benzene ring somehow spread throughout the whole ring.[137] So Hückel first had to interpret the electron configuration of the benzene ring and then explain its behavior upon multiple substitution. This "holistic" approach opened the way for Hückel not only to define the electron configuration of benzene but also to interpret by means of quantum theory the reactions

---

[134]Thus Hückel demonstrated very early on that quantum chemistry analysed not only individual molecules and substances but also classes of substances. Post-modernists and anti-reductionists often overlook this. On this issue, Cf. Primas. V., Müller-H.: *Elementare Quantenchemie*. Teubner, Stuttgart 1984, Kap.5.5., 289–302. Janich, P. und Psarros, N. (Ed.): *Die Sprache der Chemie*. Königshausen & Neumann, Würzburg, 1996; Psaros, N., Ruthenberg, K. und Schummer, J. (Ed.): *Philosophie der Chemie. Bestandsaufnahme und Ausblick*. Königshausen & Neumann, Würzburg, 1996; Janich, P. und Psarros, N. (Ed.): *The Autonomy of Chemistry. 3rd Erlenmeyer-Colloquy for the Philosophy of Chemistry*, Königshausen & Neumann, Würzburg, 1996.

[135]I note here that Hückel does not mention the previously cited paper in which Armit and Robinson introduced the electron sextet.

[136]*Chemiker im Gespräch: Erich Hückel*, p. 184.

[137]Ibid.

of a monosubstituted benzene that had been subjected to a series of substitutions after its electronic structure had been perturbed in a specific manner.[138]

The initial assumption Hückel made in his paper was that in a cyclic system with the generic formula C_nH_n, the C atoms are arranged in a regular plane polygon with n corners. Furthermore Hückel assumed on the basis of chemical structure that each carbon atom on the ring is surrounded by 7 electrons, because each carbon atom brings 6 electrons with it and each hydrogen atom 1 electron in the C–H bond. Every carbon atom has 2 electrons in its inner K shell, which do not participate in chemical bonding. Of the 5 remaining electrons, two are used up to establish the valence bond with the hydrogen atom attached to this carbon atom. Two other electrons are used up in establishing the valence bonds with the two neighboring carbon atoms on the ring. So one electron is left over for each of the carbon atoms occupying one of the n corners of the polygon. Hückel considered "the behavior of these n electrons, not assignable to localized valences afforded by electron pairs," decisive for the "binding state of the ring."[139]

Disregarding the exchanges between the force fields of neighboring carbon atoms and their electrons, Hückel employed the symmetry properties of the ring in reasoning that the "quantum state" belonging to each of the remainder n electrons is characterized by a p eigenfunction with a nodal plane in the planar ring, considered here to be *horizontal*. That is why Hückel called these *n* electrons $[p]_h$ electrons.[140] In total Hückel described for each carbon atom plus hydrogen atom on the ring the following energetically lowest state:

$$[s]^2, [p]^2, [p]_v^2, [p]_h$$

One characteristic of this arrangement is that three electron pairs are locally linked. They are the three simple bonds of each of the carbon atoms. There remains a single electron in a $[p]_h$ state that cannot be linked locally and whose eigenfunction exhibits a new symmetry. Thus there are n such $[p]_h$ electrons in the whole ring that are not assignable to any specific carbon atom or atomic pair. Hückel then examined the behavior of these n electrons within the field of the planar ring structure with paired locally bound electrons by two methods of approximation. He applied two different perturbation calculations that had to be governed by the degree of overlapping between the potential fields of neighboring carbon atoms on the ring and the intensity of the electronic interaction.

A few years previously Walter Heitler and Fritz London had developed a method for describing the formation of a hydrogen molecule from two hydrogen atoms. In it

---

138 Hückel discussed the quantum mechanics underlying the chemical behavior of substituted benzenes in a second paper. See Section 2.2.4 below.

139 Hückel, E.: Quantentheoretische Beiträge I. Die Elektronenkonfiguration des Benzols, p. 215.

140 Now they are called π electrons.

the two effective quantities are taken simultaneously into account.[141] Their starting assumption was that, to first approximation, every hydrogen atom has assigned to it a specific electron in a determined quantum state. They then took into account that the two electrons are indistinguishable. It is not possible to determine which of them would be assigned to which of the two nucleii. The exchange procedure that takes the indistinguishability between the two electrons into account in the form of what is called an exchange integral is essential for this early form of the VB method.[142] Applying methods from group theory, Werner Heisenberg used this exchange procedure for calculations involving large numbers of atoms to explain ferromagnetic effects in metals.[143] Soon afterwards the American John C. Slater developed a method – purportedly to save some colleagues from the "group plague"[144] – that Hückel praised in his article as "a practical and elegant procedure."[145] It effectively circumvented group theory by using a determinant method[146] to deal with the exchange between large numbers of atoms.[147] Felix Bloch subsequently employed Slater's method for the theory of ferromagnetism, preferring it over Heisenberg's method using group theory.[148] In his first approximative procedure, which Hückel called the "first method," he adopted the formalisms devised by Slater and Bloch for electrons in a crystal lattice.[149]

Hückel's "first method" addressed the interaction of n many $[p]_h$ electrons within the field of n many CH groups according to the procedure mentioned above, assigning for each carbon atom one $[p]_h$ electron in a specific quantum state with eigenfunction $\Phi_f(r_{if})$, with the indices indicating the i-th electron in atom f. The total eigenfunction, which incorporates the exchanges between the $[p]_h$ electrons, is described as a linear combination of products of $\Phi_f(r_{if})$, disregarding spin-orbit coupling and the interactions between spin orientations. Next, spin is taken into account and only the linear combinations obeying Pauli's exclusion principle are

---

141 Heitler, W. und London, F.: *Wechselwirkung neutraler Atome und homöopolare Bindung nach der Quantenmechanik*, in: Z P 44 (1927), 455–472.

142 Note that this exchange integral of the VB method is quite different from the exchange integral in use in the MO method to be discussed later. It is also called the resonance integral and is very different from the "exchange integral" employed in the Hartree–Fock method.

143 Heisenberg, W.: *Zur Theorie des Ferromagnetismus*, in : Z P 49 (1928), 619–636.

144 Slater, J. C.: *Solid State and Molecular Theory: A Scientific Biography*. New York, 1975, p. 62.

145 Hückel, E.: Quantentheoretische Beiträge I. Die Elektronenkonfiguration des Benzols, p. 218.

146 They are in the current references well-known as "Slater determinants".

147 Slater, J. C.: *The Theory of Complex Spectra*, in: Physical Review 34 (1929), 1293–1322; *Cohesion in Monovalent Metals*, in: Physical Review 35 (1930), 511–529. Slater originally used his method to cover the interactions between many electrons in an atom and then, in a second publication from 1930, to comprehend the interactions of electrons within a metal lattice. See Hoddeson, L., Baym, G., Eckert, M.: *The Development of the Quantum-Mechanical Electron Theory of Metals, 1928–1933*, in: Reviews of Modern Physics 59 (1987), 287–327. Schweber, S.: *The Joung John Clark Slater and the Development of Quantum Chemistry*, in: Historical Studies in the Physical and Biological Sciences 20 (1990), 339–406.

148 Bloch, F.: *Zur Theorie des Ferromagnetismus*, in: Z P 61 (1930), 206–219.

149 Hückel's "first method" is essentially a valence bond (VB) method.

applied. Hückel then employed this procedure in the form developed by Slater and Bloch for electrons in a crystal lattice. But he encountered difficulties in resolving the secular problem.

Peierls, a student of Sommerfeld in Munich who completed his dissertation on the thermal conductivity of solids under Heisenberg at Leipzig in 1929, offered assistance to Hückel in solving this problem. In 1930 Peierl was Pauli's assistant in Zurich, where he was working on another application of quantum mechanics to the electron theory of metals for his habilitation thesis.[150] In a letter to Hückel at the end of June, Peierl suggested a possibile solution to the problem, making the following consideration: "I send you in the enclosed the calculations we recently made on the blackboard. [...] But it seems to me that it doesn't work simply. In your case it should be possible, however, to reduce the secular determinant by this method in a simple way."[151] Hückel decided not to take up Peierls's suggested method.[152] Hückel's solution to the secular problem employs the ring's cyclic symmetry. Because of this symmetry, the set of all terms (or eigenfunctions and their corresponding eigenvalues) breaks down into level systems that can be characterized by a number $K = 0, 1, ..., n-1$. Their mirror-image qualities and the values of the resultant spins must be characterized in addition. Hückel calculated out the energetic series of eigenvalues (electron states), which could be expressed as a multiple of the exchange integral J, not only for benzene ($n = 6$), but also for the cases $N = 3$, 4 and 5.[153] Finally, he compared his results against those he obtained from the next method.[154]

Let me note here that the "first method" Hückel discussed was only applicable to cases in which the number N "of electrons not attributable to unique bonds" concurred with the number of atoms n on the ring: N = n. Therefore, it was not applicable to other aromatic compounds like thiophene, pyrrole, etc., because $N \neq n$. But Hückel still thought it "necessary to follow the first method through to the end as well, in order to ascertain to what extent its results agreed with the second method, and in what way they deviated."[155]

Hückel's other approximative method, his "second method," is an MO procedure and is known today as the HMO method (Hückel's molecular orbital method).[156]

[150]Hoddeson, L., Baym, G., Eckert, M.: *The Development of the Quantum-Mechanical Electron Theory of Metals, 1928–1933*, in: Reviews of Modern Physics 59 (1987), 293–296.

[151]SBPK, Papers of Hückel, Box 6, Folder 5.116, Letter Peierls to Hückel, Zürich, 28.6.1930.

[152]Hückel explained his decision in a footnote of his publication. See Hückel, E: *Quantentheoretische Beiträge I. Die Elektronenkonfiguration des Benzols.*, 225, footnote 1.

[153]Ibid., pp. 272–286.

[154]Ibid., pp. 248–258.

[155]Hückel, E.: Quantentheoretische Beiträge I. Die Elektronenkonfiguration des Benzols, p. 220.

[156]It is used and explained in a wide variety of textbooks. Cf. Heilbronner, E., Bock, H.: *Das HMO-Modell und seine Anwendung*. Verlag Chemie, GmbH, Weinheim, 1968; Coulson, C. A., O'Leary, B., Mallion, R. B.: *Hückel Theory for Organic Chemists*. Academic Press, London, 1978; Dewar, M. J. S.: *The Molecular Orbital Theory of Organic Molecules*. McGraw-Hill, New York

Hückel's "second method" essentially adopted a theory developed by Bloch in 1928 at Leipzig to explain phenomena like the electric conductivity of a metal lattice.[157] After completing his studies in physics under Erwin Schrödinger in Zurich, Bloch left for Leipzig in the winter term of 1927–28 at the advice of Debye to work on his doctorate under Heisenberg.[158] His thesis treated the topic of the paper referred to above: "On the quantum mechanics of electrons in crystal lattices." It dealt with a problem left open by Sommerfeld's electron theory of metals, namely, the states of the electrons within the energy field of atoms forming a crystal lattice.[159]

Ignoring the mutual interactions, Bloch treated the motion of the electrons in the crystal lattice not as free but influenced by a field of force of the same periodicity as the crystal's lattice structure.[160] Following Schrödinger's interpretation, he also described the electrons physically as "de Broglie waves [...] that are modulated to the rhythm of the crystal's structure."[161]

Hückel retraced Bloch's considerations in his analysis of the quantum states of a single $[p]_h$ electron in a field of force of the same periodicity as the structure of the cyclic compound. This field of force is caused by the structure and the other $[p]_h$ electrons beside the one under consideration.[162] So one could imagine Hückel's benzene lattice as composed of regular hexagons of $[p]_h$ electrons of carbon.

For calculation purposes Hückel introduced cylindrical coordinates (r, z, φ) for the regular ring of n corners with the center of the n-polygon as the origin. The potential for individual electrons was a periodic function of φ with the period $2\pi/n$:

$$V(r,z,\varphi) = V\left(r,z,\varphi + \frac{2\pi}{n}g\right); \quad g = 1, 2, \ldots, n-1$$

1969; Pullman, B. and Pullman, A.: *Les théories électroniques de la chimie organique*. Masson, Paris, 1952; Yates, K.: *Hückel Molecular Orbital Theory*. Academic Press, New York, 1978 Kutzelnigg, W.: *Einführung in die Theoretische Chemie*. Band 2: *Die chemische Bindung*. VCH, Weinheim, New York, 1978; Scholz, M. und Köhler, H.: *Quantenchemie – Quantenchemische Näherungsverfahren und ihre Anwendung in der organischen Chemie*. VEB Deutscher Verlag der Wissenschaften, Berlin 1981; Streitweiser, A.: *Molecular Orbital Theory for Organic Chemists*. Wiley, New York, 1961.

[157]Bloch, F.: *Über die Quantenmechanik der Elektronen in Kristallgittern*, in: ZP 52 (1928), 555–600.

[158]Bloch, F.: Reminiscences of Heisenberg and the early days of quantum mechanics (1976).

[159]The paper by Bloch mentioned above and subsequent papers and survey articles also by Rudolf Peierls and other American, German and Russian physicists appearing up to 1934 formed the theoretical foundations of modern solid state physics. See Eckert, M.: *Die Atomphysiker*, Kap.6, p. 133.

[160]Bloch, F.: Über die Quantenmechanik der Elektronen in Kristallgittern, pp. 555–556.

[161]Ibid. 559.

[162]Hückel, E: Quantentheoretische Beiträge I. Die Elektronenkonfiguration des Benzols, p. 240.

The corresponding Schrödinger equation for each electron is:

$$\Delta\chi + \frac{8\pi^2 m}{h^2}\left[W - V(r,z,\varphi)\right]\chi = 0$$

As concerns the analytic form of the eigenfunction, Hückel again relied on Bloch's conclusions. In the paper mentioned above, Bloch had demonstrated by means of group theory that the eigenfunction of an electron could be described as the product of a spatially periodic function $u^k(r, z, \phi)$ and a freely propagating wave:[163]

$$\chi^k(r,z,\varphi) = e^{\iota k\varphi} u^k(r,z,\varphi)\,; \quad k = 0, \pm 1, \pm 2, \ldots$$

where $u^k(r, z, \phi)$ is a ϕ periodic function with the period $2\pi/n$ and k is either a positive or negative whole number or zero. This function depends on the development of the potential in each case. Consequently, the potential V determines the specific form of the eigenfunctions χ^k along with their corresponding eigenvalues. The potential is composed of the overlapping potentials of the individual atoms of the cyclic molecule and the other electrons surrounding them. So, in principle, it depends on the given molecule's structural chemical formula (of the ring-shaped frame).

As concerns the physical interpretation of the eigenfunctions χ^k, Hückel, like Bloch, adopted Schrödinger's visual interpretation, imagining them as "circulatory electron waves that as a consequence of the periodicity in ϕ are modulated at this periodicity."[164]

For the calculation of the eigenvalues and eigenfunctions Hückel employed an approximative procedure indicated by Bloch in which the overlapping fields of potential of the individual atoms are treated as perturbations. Bloch's method yielded the following result for eigenvalue W:

$$W^k = \alpha + 2\beta\cos\left(\frac{2\pi k}{n}\right)$$

α and β are two parameters resulting from integrals over the values contributed by the individual atoms. They have the following analytic form:

$$\alpha = \iiint \left[\mathbf{V} - \mathbf{U}_f\right] \Psi_f^{0^2}\, r\, dr\, dz\, d\varphi$$

$$\beta = \iiint \left[\mathbf{V} - \mathbf{U}_f\right] \Psi_f^0\, \Psi_{f+1}^0\, r\, dr\, dz\, d\varphi$$

[163] Bloch, F.: Über die Quantenmechanik der Elektronen in Kristallgittern, p. 559.

[164] Hückel, E.: Quantentheoretische Beiträge I. Die Elektronenkonfiguration des Benzols, p. 241.

V describes the true development of the "Hartree potential"[165] and $U_f(r, z, \phi)$ the theoretical development of the f-th atom by itself. $\Psi^0{}_f(r, z, \phi)$ is the eigenfunction of the lowest eigenvalue of a $[p]_h$ electron in the theoretical potential U_f. The quantity α represents the potential energy of the unperturbed charge distribution of the $[p]_h$ electron located with the individual atom in the fields of the neighboring atoms. Since α describes an electron bound to an atom, α is negative. The quantity β describes the quantum-mechanical "resonance interaction" between the two neighboring atomic eigenfunctions (ψ_f^0 and ψ_{f+1}^0). It is the stabilizing energy of an electron relative to α, when two such entities are interacting within a certain distance. Hückel showed that β is likewise negative for $[p]_h$ electrons.

The eigenfunction for eigenvalue W^k is:

$$\chi^k(r, z, \varphi) = \frac{1}{\sqrt{n}} \sum_{f=0}^{n-1} \varepsilon_n^{fk} \Psi_f^0(\varphi)$$

The eigenvalues and eigenfunctions are characterized by the following values for k: $k = 0, \pm1, \pm2, ..., \pm n/2$, where n is even. Such "quantum numbers" characterize the electron states in their energetic sequence. Hückel then plugged in the electron states, taking resonance effects and Pauli's exclusion principle into account, in a like manner to the MO method by Hund and Mulliken for diatomic molecules.[166] Drastic negligence of the "interactions between the spins and electronic motions," the energy exchanges between electrons in k states and skillful use of molecular symmetry led Hückel to the following general finding:

> There consequently results for an n ring the electronic states characterized by the k "quantum numbers." k = 0 yields (without spin) one state, k = 1 is doubly degenerate and yields (without spin) two states, k = 2 likewise, etc. [...] Because, according to the Pauli principle, each state can only be occupied doubly, one obtains a first complete electron shell for 2 electrons, a second for 4 additional electrons (2 + 4 = 6) and another for 4 more electrons (2 + 4 + 4 = 10).[167]

These results, originating from Schrödinger's differential equation, led Hückel to a quantum-theoretical interpretation of aromaticity: The numbers 2, 6 and 10

---

[165]Like Bloch, Hückel neglected to first approximation the exchange of energy between the electrons and first determined the quantum states of an electron in a field of force of the periodicity of the ring by the Hartree method.

[166]Hund, F.: Zur Deutung einiger Erscheinungen in den Molekelspektren, in: ZP 36 (1926), 657–674; Zur Deutung der Molekelspektren. I, in: ZP 40 (1927), 742–764; Zur Deutung der Molekelspektren. II, in: ZP 42 (1927), 93–120; Zur Deutung der Molekelspektren. IV, in: ZP 51 (1928), 759–795; *Chemical Binding*, in: Transactions of the Faraday Society 25 (1929), 646–648; Mulliken, R. S.: *The Assignment of Quantum Numbers for Electrons in Molecules. I*, in: Physical Review 32 (1928), 186–222; *The Assignment of Quantum Numbers for Electrons in Molecules. II. Correlation of Molecular and Atomic Electron States*, in: Physical Review 32 (1928), 761–772; *The Assignment of Quantum Numbers for Electrons in Molecules. III. Diatomic Hydrides*, in: Physical Review 33 (1929), 730–747; *The Interpretation of Band Spectra. I, II, III*, in: Reviews of Modern Physics 2 (1930), 60–115, 3 (1931), 89–155, 4 (1932), 1–86.

[167]Hückel, E.: Quantentheoretische Beiträge I. Die Elektronenkonfiguration des Benzols, p. 255.

describe "complete electron shells," which endow monocyclic aromatic molecules with particular stability. *The occurrence of the numbers 2, 6 and 10 for "complete electron shells" in a quantum-mechanical analysis of cyclic compounds is characteristic of such compounds.* They resulted only from the "second method," not from the "first."[168]

Today chemists still employ the "Hückel rule" of the form: "4n + 2," where n = 0, 1, 2, 3, etc., to determine whether a given organic cyclic compound with 4n + 2 π-electrons is classifiable among the aromatic compounds.[169]

2.2.2.2 Conclusions

But coming back to Hückel, what were the cornerstones of his quantum-mechanical theory of aromaticity?

First, Hückel emphasized that it was not the number of atoms forming the ring but the number of electrons forming a "complete electron shell" that determined aromaticity.

Second, as he pointed out, such "complete electron shells" explained the chemical stability of aromatic cyclic systems, in particular, their strong resistance to chemical addition. Among the aromatic species, hence ones with "complete electron shells," only the lowest energy states are occupied. That is why they are particularly low in energy and as a consequence stable and especially chemically resilient. In Hückel's mind, the aromatic molecule constituted the model example of a chemically stable species.

Hückel thus offered a chemical explanation, not a physical one, for a molecule's energetic stability. What is the difference between the two explanations?

Physicists and chemists have different understandings of the term "molecule." In physics every energetically stable aggregate of atoms is called a molecule. In chemistry the term molecule is only used for clusters of atoms of relatively long duration in "a normal chemical environment" permitting selection of the molecule involved from among other kinds of molecules for specific classification. Hence, a chemist's definition of a molecule is not based solely on a strong resistance to chemical action with ubiquitous reagents. All other atomic arrangements are referred to as "radicals." The cause of a molecule's reactivity is, among other things, its electronic structure constituting shells of greater or lesser degrees of completion. "Complete electron shells" mean, quantum-theoretically speaking, that the total angular momentum of the electrons equals zero and that all states are occupied by two electrons of antiparallel spins. Hückel explicates these facts in his paper:

---

[168]This is based on the fact that Hückel's "second method" yields from k = 0 the state of lowest energy, which is not degenerate, whereas for k = 1 the corresponding state is doubly degenerate, i.e., it appears as a pair of equal energy. I also point out here that Hückel's "first method" yielded fewer states than the "second method."

[169]This brief formulation of the Hückel rule as "4n + 2" was probably first used by E. Doering and F. L. Detert in 1951. See Berson, J. A.: *Erich Hückel – Pionier der Organischen Quantenchemie* (1996), p. 2932.

As already mentioned, the energy content furthermore does not by any means alone govern *the stability of a compound in a chemical sense.* In this the reactivity of a compound is also decisive. This reactivity depends, among other things, on how the energy responds to a disturbance in the atomic arrangement (changes in the separating distances), how easily the molecule is excited, how easily it takes on electrons, etc. In general, much experimental data has been gathered about the correspondence between the constitution and reactivity of organic molecules. Yet only a very modest number of satisfactory theoretical conceptions about it exist. We now believe we can contribute a new aspect in the case of cyclic compounds considered here. The introduction alluded to the importance ascribed to the number 6 for "double-bond electrons" in chemistry and stated that *in a certain sense, this number 6* corresponded to *a complete electron shell.*[170]

Third, Hückel himself later[171] referred to the analogy between the atomic stability of the noble gases and aromatic molecules. The unusual stability of both species is caused by "complete electron shells." In the case of the noble gases, the valence electron octet performs the key role.

Fourth, Hückel extended his theory for benzene to other aromatic species and was able to predict the existence of other forms. In fact, for $n = 6$ members to a ring corresponding to benzene, Hückel obtained 6 electron states in the ground state with a complete electron shell formed of 6 electrons. According to Hückel's description, the electron states of benzene are occupied as follows: 001111. Figure 2.9 illustrates his result.

For $n = 4$ and $n = 8$ members of a ring, such as for cyclobutadiene (at that time yet to be sythesized) or cycloalkene, Hückel did not obtain complete electron shells from his "second method," because their electron states are occupied as follows: 0011 and 00111122 (Fig. 2.10).

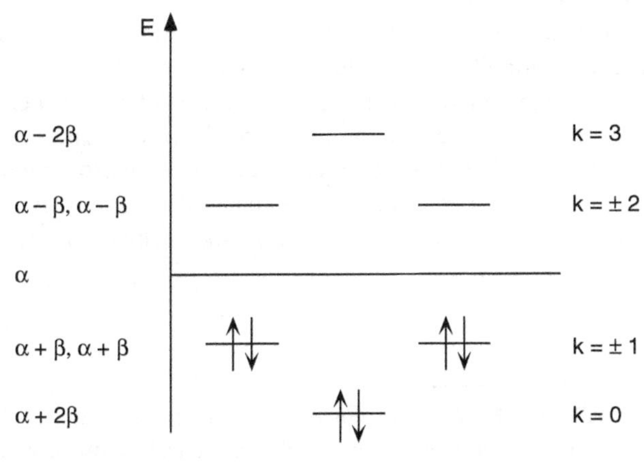

Fig. 2.9 The 6 one-electron states of benzene and how they are occupied in the ground state

[170] Hückel, E.: Quantentheoretische Beiträge I. Die Elektronenkonfiguration des Benzols, pp. 254–255. (Emphasis mine).

[171] Hückel, E.: *Grundzüge der Theorie ungesättigter und aromatischer Verbindungen*, in: Z. Elektrochem. 43 (1937), 752–788, 827–849, p. 778.

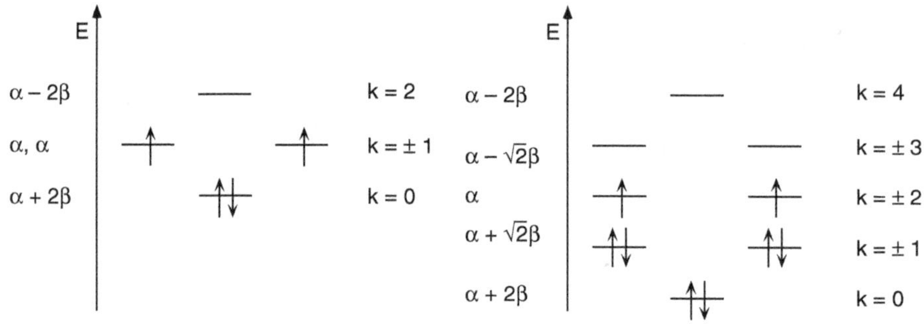

Fig. 2.10 Electron states of cyclobutadiene and cycloalkene (cyclo-octatetraene)

According to Hückel's interpretion, this signified that these cyclic systems are not aromatic in character and are more reactive than benzene, as Willstätter and Waster had already demonstrated with cycloalkene.[172] This, Hückel emphasized, was only valid under the condition that the nonplanar arrangement of the ring was unable to have a major influence on its stability and chemical properties.[173]

Thus Hückel managed to solve the enigma for classical structural theory in organic chemistry posed by the differing chemical behaviors of benzene, cyclobutadiene and cycloalkene. Moreover, Hückel's quantum-mechanical approach to aromaticity was able to predict the existence of a 10-membered ring ([10] annulene) of low reactivity from its closed electron shell: 0011112222. His explicit prediction inspired its eventual synthesis: "It would therefore be interesting to try to produce this compound, and if it worked, to watch whether, unlike the 8-membered ring, it manifests a more aromatic character."[174]

Hückel's accomplishment went beyond a quantum-theoretical underpinning of the aromatic electron sextet that Armit and Robinson had formulated. It provided a first firm foundation for a quantum-mechanical theory of aromaticity, offering a criterion for the assignment of a given substance among the class of aromatic compounds. His theoretical considerations on aromaticity also enabled quantitative comparisons of the energies of the molecules under consideration. Nevertheless, the simplifications introduced in Hückel's method have a detrimental effect on its

[172]Willstätter, R., Waser, E.: *Über Cyclo-octatetraen*, in: B. 44 (1911), 3423–3445.

[173]Hückel, E.: Quantentheoretische Beiträge I. Die Elektronenkonfiguration des Benzols, p. 256.

[174]Ibid., 255–256. Experimental confirmation of Hückel's prediction in fact had to await the end of World War II, after which [10] annulene and other cyclic polyenes (annulenes) of the same aromatic character were synthesized. Hückel's preliminary results on the stability of as yet unknown aromatic compounds triggered research specifically focusing on the synthesis of new substances of that class. See Garratt, J. P.: *Aromaticity*. John Wiley & Sons, New York, 1986, Chap. 4 The Annulenes.

quantitative predictions about the series of energy states.[175] Heisenberg also pointed this out in a letter to Hückel from April 10, 1931, commenting:

> There are, of course, plenty of more or less arbitrary assumptions inside your calculations, but I think one can justify them quite well from the viewpoint of a consistent Hartree method. [...] So, qualitatively speaking, I think that your considerations are satisfactory but one naturally cannot place too much value on it quantitatively.[176]

Undoubtedly, Hückel was aware of the qualitative nature of his considerations. However, the fact that the results obtained by disregarding various interactions agreed with existing chemical experience rather confirmed the approximative validity of his assumptions.[177]

Fifth, symmetry plays as fundamental a role as the principle of "complete electron shells" in Hückel's theory of aromaticity. The one-electron states (orbitals) are first classified by their symmetry. Then their numbers are evaluated according to the coefficient with which the parameter β enters into the energy expression. β denotes the resonance interaction between neighboring atoms. The energy expression delivers the energetic sequence of states, for which the following applies: The more nodes the relevant eigenfunctions have, the higher is their energy.[178] The symmetries of the electron configurations, i.e., the occupied orbitals, obviously could not be depicted in normal three-dimensional space. Hückel was strident in his criticism of visual representations of the benzene molecule by means of two Kekulé-style structural formulas:

> The present interpretation of the electron configuration of benzene strips Kekulé's formulation of the benzene ring with alternating double and single bonds of its meaning. [...] [In the case of benzene] all 6 atoms are completely equivalent, not just in space but also in configurational space. Furthermore, the eigenfunction (not just for the spatial distribution of charges) is symmetric with respect to the mirror images of all the molecule's planes of symmetry. (This applies, incidentally, irrespective of our assumption that we are dealing with $[p]_h$ electrons.)
>
> [...] One thus sees that the symmetries of the electron configurations cannot be represented in normal space and that it must therefore also be impossible to draw a distinction between

[175]The greatest deficiencies of Hückel's method lay in the rather heavy-handed approximations, preventing any reliable statements about such quantities as resonance energies or electronic spectra. That is why a number of other procedures were developed, particularly after World War II, to be able to make more accurate quantitative predictions about the aromaticity of compounds so as to arrive at a better understanding of their unusual energetic stability. See Garratt, P., Vollhardt, P.: *Aromatizität*, Stuttgart, 1973.

[176]SBPK, Papers of Hückel, Box 6, Folder 5.16, Letter Heisenberg to Hückel, 10. April 1931.

[177]Hückel, E.: Quantentheoretishe Beiträge I. Die Elektronenkonfiguration des Benzols, p. 270.

[178]This was one of the valuable results of Hückel's quantum theory of aromaticity. Andrew Streitwieser, one of Hückel's postwar advocates in America, emphasizes in his autobiographical notes: "The real value of Hückel Theory is that it gives the correct nodal properties of MOs, and these alone can lead to important predictions and understanding of chemistry." See Streitwieser, A.: *A Lifetime of Synergy with Theory and Experiment*. American Chemical Society, Washington, DC, 1997.

the binding states in the various cases by means of such spatial symbols as localized double and single bonds, even if they are regarded as "oscillatory" or "fluid."[179]

Sixth, Hückel's quantum theoretical contribution evidently carried his critical proof of classical structural depictions of the organic molecule a step beyond his quantum theory of double bonding from the foregoing year. Hückel pointed out that Kekulé's formulas for benzene did not reflect real structures but only conventions or representational consensus. Hückel's conception of the benzene molecule, on the other hand, was a purely mathematical abstraction in configurational space, with the molecular symmetry as an important prerequisite. It, too, was a conventional representation of the chemical facts. Nevertheless, Hückel was convinced that experimental observation legitimized it much better than valence line diagrams. He also discussed these ideas with Max Born. In a letter from November 29, 1931 he outlined his methodological preference for the new quantum-theoretical valence diagram:

The difference between the valence line diagram and the valence diagram according to quantum theory appears to me to provide a nice illustration and a decision in favor of the quantum-theoretical picture of the situation with aromatic compounds. For instance, the line diagram of benzene allows a prediction of both possibilities. [...] But in quantum theory only one possibility exists. There, in configurational space, all 6 atoms are equivalent. This equivalence cannot be achieved with the line diagram. But it is required, not just by quantum theory but also by experiment. Consequently, this is a case where experiment decides in favor of the quantum-theoretical diagram and against the line diagram.[180]

Hückel regarded Kekulé's valence line diagram merely as a symbolic representation that failed to do adequate justice to the quantum-mechanical forces and resonance interactions between the electrons and carbon atoms. For this reason a full understanding of the valence line diagram was not possible without a more basic foundation afforded by quantum theory. Its abstract configurational space could better describe the experimentally confirmed equivalence of the six carbon atoms. Thus the quantum-mechanical diagram reflected a deeper conceptual level. This is a fine example of what Hilbert generally referred to as a "relaying of the foundations deeper down,"[181] and it is perhaps not going too far to discover here a hint of Hilbert's influence on Erich Hückel's theoretical conceptions of science.

Hückel did not limit the application of his "second method" to cyclic systems in which "the number of electrons are not assignable to simple pairs of bonds" agreeing with the number of atoms on the ring. He also applied it to charged monocyclic polyenes (monocyclic ions).[182] He was able to explain, for example, why cyclopentadiene (C_5H_6) – unlike cycloheptatriene (C_7H_8) – forms a stable potassium salt.

[179]Hückel, E.: Quantentheoretische Beiträge I. Die Elektronenkonfiguration des Benzols, 256–257.

[180]SBPK, Papers of Hückel, Box 6, Folder 5.11, Letter Hückel to Born, Cannstatt (Stuttgart) 29. November 1931.

[181]See Section 1.7.1 above.

[182]Hückel, E.: Quantentheoretische Beiträge I. Die Elektronenkonfiguration des Benzols, p. 257 f.

This salt demonstrated that the $C_5H_6^-$ ion had to be quite stable. Thus one finds a 5-membered ring with 6 electrons occupied, like the 6-membered ring of benzene, as 001111. Hückel interpreted this "as a tendency toward completion of the closed 6-membered electron shell."[183] By contrast, the $C_7H_7^-$ ion would be occupied in the same way as cyclo-octatetraene (00111122), i.e., lacking a complete outer electron shell. Hückel expected cyclononatetraene (C_9H_{10}) to form a stable $C_9H_9^-$ ion again, because it could form a complete electron group of 10 electrons with occupation numbers 0011112222.

Hückel's theory thus clarified many long familiar peculiarities of aromatic compounds observed in organic chemistry, even explaining the anomalous properties of annulenes[184] and monocyclic charged polyenes with 4n + 2 $[p]_h$-electrons. His theoretical considerations about the molecular stability of as yet unknown compounds also served as a starting point for research concentrating on the synthesis of new substances.[185] Until the end of World War II, the main advances in the field of aromaticity were rather of a theoretical nature. It was only afterwards that various research teams synthesized the monocyclic aromatic compounds predicted by Hückel's theory.[186]

The team of researchers led by Franz Sondheimer is a prime example. They started working on the synthesis of a series of aromatic annulenes in 1956.[187] As Berson pointed out, the most persuasive proof of acknowledgment of Hückel's ideas about aromaticity appear to be the carbo-cations tropeoline and cyclopropenyl, synthesized in the 1950s and 1960s.[188] In 1962 Breslow confirmed, after much effort and using great synthetic resourcefulness, the Hückel rule for $n = 0$ by producing the cyclopropenyl cation (Fig. 2.11).

---

[183] Ibid., 257.

[184] Cyclobutadienes, benzenes and cyclo-octatetraenes (cycloalkenes) were regarded after 1945 as members of a homologous series of conjugate, monocyclic carbon atoms of the general form $(C_2H_2)_n$, where the index n denotes the size of the ring. Such systems were called *annulenes*. According to this general nomenclature, benzene is referred to as a [6] annulene and cyclo-octatetraene as an [8] annulene.

[185] During the 1980s the first qualitative statements about the aromaticity of fullerenes were made within the framework of Hückel's theory, triggering intensive research on this problem. Minkin, V. I., Glukhovtsev, M. N., Simkin, B. Y. A.: *Aromaticity and Antiaromaticity*, New York, 1994; Haymet, A. D. J.: *C_{120} and C_{60}: Archimedean Solids Constructed from sp^2 Hybridized Carbon Atoms*, in: Chemical Physics Letters 122 (1985), 421–424. Sobczyñska, D.: Fullerenes: *The Philosophical Aspects of Their Discovery*, in: *ARS MUTANDI. Issues in Philosophy and History of Chemistry*, edited by Nikos Psarros and Kostas Gavroglu. Leipziger Universitätsverlag, Leipzig, 1999.

[186] Cf. *Theoretical Organic Chemistry. Proceedings and Discussions of the Kekulé Symposium.* Butterworths Scientific Publications, London, 1959; *Aromaticity. An International Symposium.* The Chemical Society Special Publication No. 21, London, 1967; Garratt, P. und Vollhardt, P.: *Aromatizität.* Georg Thieme Verlag, Stuttgart, 1973, Chap. 4.

[187] Cf. Garratt, P. und Vollhardt, P.: *Aromatizität,* Chap. 3. Die Annulene.

[188] Berson, J. A.: *Chemical Creativity. Ideas from the Work of Woodward, Hückel, Meerwein, and Others.* Wiley-VCH, Weinheim, 1998, p. 53; Erich Hückel – Pionier der Organischen Quantenchemie, p. 2932.

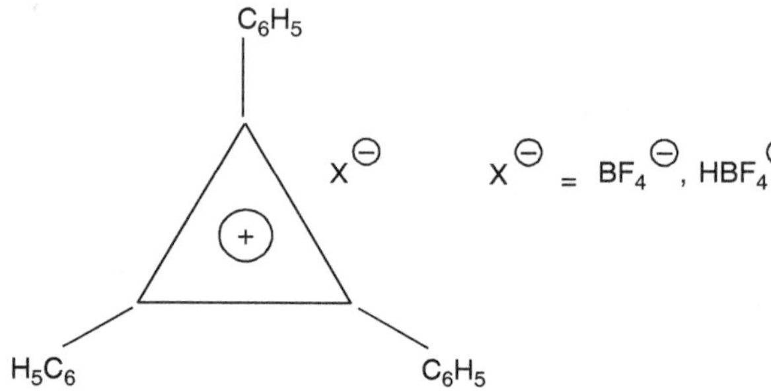

Fig. 2.11 Cyclopropenyl cation

It is a three-membered monocyclic ring with a cyclic bond between the three centers with one delocalized positive charge on the three carbon atoms.

2.2.3 Hückel's Model of Benzene and Friedrich Hund's Localization Conditions

In his paper on the quantum-mechanical interpretation of benzene's electron configuration, Hückel used qualitative arguments based on symmetry considerations to deny that benzene could be represented as alternating localized double and single bonds. The root of this argument lay in the concept of localized and nonlocalized bonds. But Hückel offered no solution by quantum theory to the basic question of when it was permissible to assume a localized bond between two electrons depicted by chemists as a valence line.

Born pointed out this deficiency to Hückel in a letter from December 5, 1931 after he and members of his seminar at Göttingen had gone over Hückel's quantum-theoretical valence diagram.

> We discussed your last paper during one of our last seminar sessions. [...] Your basic idea that a certain group of electrons cannot be assigned to one of the valence lines appeared quite plausible to us, although not compellingly convincing. What I am missing is an assignment intermediate to the chemical symbol for the double line [...]. I unfortunately cannot continue to work on these issues now, because I want to finish up my book on optics.[189]

That same year Friedrich Hund in Leipzig picked up where Born had left off. He and Heisenberg had also been exchanging thoughts with Hückel.

After studying Hückel's paper on the electron configuration of benzene, Hund sent him in a letter dated May 16, 1931 a few comments about Hückel's method for representing the energetic ground state of each carbon atom of benzene.[190] Hund indicated to Hückel that his formulations $[s]^2$, $[p]^2$, $[p]^2{}_v$, $[p]_h$ were not easily understandable to the reader. Hund suggested the following alternative: $[1s]^2$

[189] SBPK, Papers of Hückel, Box 6, Folder 5.11, Letter Born to Hückel, Göttingen, 5. Dezember 1931.

[190] SBPK, Papers of Hückel, Box 6, Folder 5.17, Letter Hund to Hückel, Leipzig, 16. Mai 1931.

$[2s]^2$ [p] $[p]_v$ $[p]_h$, where [p] and $[p]_v$ electrons could be used for a localization of the C–C bond.[191] Hückel stood by his own designations, however, presumably improving his explications of them in the final version of his paper. At the same time he asked Hund to specify which parts of his paper were difficult to understand. Hund's reply was pointedly polite: "That wouldn't really be possible (besides what has already been mentioned). I wouldn't like to meddle into your business now, either; especially considering that you have a much better understanding of benzene than I."[192]

In his correspondence with Hückel, Hund also mentioned his own considerations on the interpretation of chemical bonding that he had published in two articles in 1931, soon after Hückel's paper on benzene and its aromatic derivatives appeared in *Zeitschrift für Physik*.[193] Particularly the second of these two papers explored whether the quantum-theoretically based valence diagrams by Hückel and others (Slater, Pauling, and other theoreticians) concurred with the empirically based system of rules about valence in organic chemistry. He began with the question Hückel had raised about when one was permitted to assume localized bonds. Hund accepted "localization" in general, or at least "approximate localization," "when enough eigenfunctions of the individual electrons in each atom come into consideration for bonding and enough electrons for the electron eigenfunctions of the molecules."[194] Hund also pointed out explicitly that, under certain assumptions about the energetic sequence of σ and π bonds,[195] the most important case of "nonlocalized bonds" occurs "when there are too few available electrons to supply two electrons for each of the possible bonds in the given or approximated arrangement of atoms."[196] According to Hund, systems with nonlocalized bonds include: most crystal lattices in a solid; "aromatic cyclic compounds in organic chemistry"; and "polycycles" (naphthaline, anthracene), which "are evidently planar structures with nonlocalized π bonds, localized σ bonds and angles of exactly 120°."[197]

Hund supplemented his analyses on localized and nonlocalized bonds with a new paper received by the editors of *Zeitschrift für Physik*, on December 10, 1931.[198] He informed Hückel in a letter early the following year: "I wrote up a few things on the

---

[191] Ibid.

[192] Ibid.

[193] Hund, F.: Zur Frage der chemischen Bindung, in: ZP 73 (1931), 1–30; Zur Frage der chemischen Bindung. II. Zum Verständnis der organischen Chemie, in: ZP 73 (1931), 565–577.

[194] Ibid., p. 569.

[195] Hund assumed that the σ bond was energetically more efficient than the π bond. This is because the σ eigenfunction in the region directed opposite to the other partner has a larger value than the π eigenfunction.

[196] Hund, F.: Zur Frage der chemischen Bindung. II. Zum Verständnis der organischen Chemie, in: ZP 73 (1931), 565–577, p. 574.

[197] Ibid., pp. 574–575.

[198] Hund, F.: *Zur Theorie der schwerflüchtigen nichtleitenden Atomgitter*, in: ZP 74 (1932), 1–17. Friedrich Hund's localization conditions are still valid to day. Kutzelnigg, W.: *Friedrich Hund und die Chemie*, in: Angewandte Chemie 108 (1996), 629–643.

localization of bonds in a second paper that should be appearing shortly. The equivalence and tetrahedral structure of the 4 valences of C is, of course, immediately given, if one regards s and p as approximately degenerate."[199] In this new paper Hund formulated a few conditions governing the possibility of a localized bond in the chemical sense of the word, i.e., a bond between two centers. These conditions for localization are: the number n of available valence electrons per atom must equal the number r of bound neighboring atoms as well as the number t of "electron eigenfunctions" (valence atomic orbitals) participating in the bond for each atom. Therefore, $n = r = t$.[200] Hence, methane (CH_4) and diamond are examples of systems that can be described by localized bonds: for each carbon atom the three numbers (n, r, t) are equal to 4. Benzene, by contrast, is a typical case in which Hund's localization conditions are not satisfied. The three numbers are all equal to three for the σ bonds of the cyclic frame C_6H_6, so in the chemical sense it was still conceivable to refer to localized σ bonds (Fig. 2.12).

However, the number of neighboring atoms for the π bonds between the six available $[p]_h$ electrons was $r = 2$, while only one $[p]_h$ electron and one electron eigenfunction (valence orbital) were available ($r = 2 > n = t = 1$). Consequently, according to Hund no description was possible by means of localized bonds[201] corresponding to the chemist's valence-line diagram.[202]

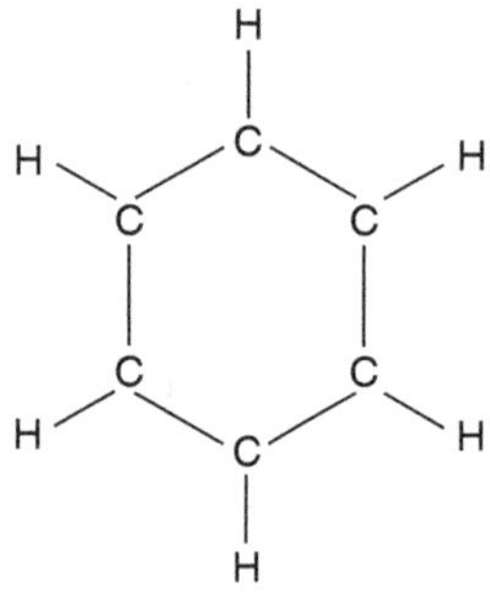

Fig. 2.12 The localized σ bonds of the benzene ring C_6H_6

[199] SBPK, Papers of Hückel, Box 6, Folder 5.17, Letter Hund to Hückel, Leipzig, 8. January, 1932.

[200] Hund, F.: *Zur Theorie der schwerflüchtigen nichtleitenden Atomgitter*, in: ZP 74 (1932), 1–17, p. 2.

[201] I point out here that for Hund "localizability" means localizability between two centers. It is also noteworthy that according to Klaus Ruedenberg the transformation of the canonical Hückel MOs to "localized MOs" leads to three-centered π orbitals. Cf. England, W., Salmon, L. S., and Ruedenberg, K. *Localized Molecular Orbitals: A Bridge Between Chemical Intuition and Molecular Quantum Mechanics*, in: Fortschritte der chemischen Forschung 23 (1971), 31–122.

[202] Hund's localization condition for covalently bound molecules is necessary but not sufficient. This means that only in cases where this condition is not satisfied can one be certain that a transformation to localized orbitals does not work. Cf. Kutzelnigg, W.: *Einführung in die Theoretische Chemie*. Band 2. VCH, Weinheim 1994, Chap. 10.

In his papers on the theory of chemical binding and crystal lattices, it was not Hund's primary purpose to calculate out particular quantities but to find a physical interpretation for familiar chemical concepts such as valence number, valence line, valence angle, etc.[203] He considered it important to assign precise quantum-mechanical terms to the fundamental concepts and rules of chemistry – in particular organic chemistry – that chemists used to sort their experimental data. This precise nomenclature was supposed to transform such expressions as "valence line," "aromatic bond," etc. into exact concepts with clearly defined ranges of applicability.[204]

In an article received by the editor's office of the *Zeitschrift für Physik* on April 29, 1932, Hückel extended his "second method" to further systems: to fused ring systems (naphthaline, anthracene and phenanthrene) as well as to diphenyl and unsaturated open chains (conjugated systems) of the form C_nH_{n+2} with both even and odd numbers of members.[205] It carried Hund's papers just mentioned a step further. Hund's localization conditions had placed Hückel's more qualitatively based considerations within a more general context. The purpose of Hückel's new article was to employ "finer gradations" of physical concepts to describe the chemical behaviors of these compounds.[206]

In discussing the results on the electron configurations of the new systems in their ground states as they related to chemical behavior, Hückel wrote:

> A *localization* of the bonds induced by the $[p]_h$ electrons is not possible. Correspondingly, the fused ring systems and diphenyl also have the highest symmetries in their ground states, in configurational just as in real space, which is congruent with the atomic arrangement.[207] [...] A *localization* of the "double bonds," for instance for naphthaline, corresponding to the figures
>
> 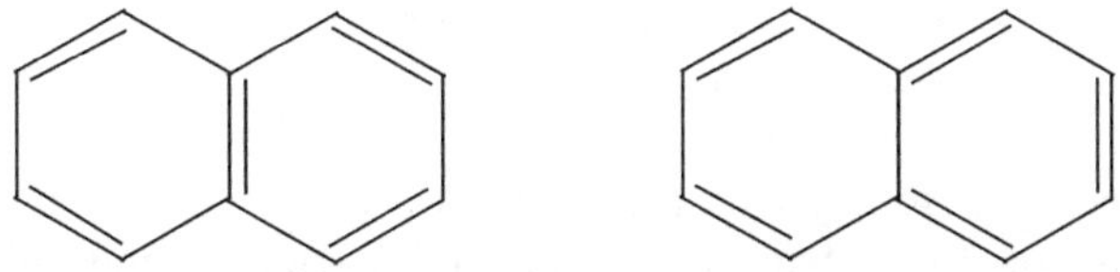
>
> is therefore not possible here either, in conformance with the chemical behavior.[208]

Hückels quantum-mechanical interpretation of fused ring systems thus led him to the conclusion that the electron configurations of such systems could only be

[203] Friedrich Hund, conversation with A. Karachalios. Göttingen August 22, 1994.

[204] Hund, F.: *Molekelmodelle*, in: *Das Molekül und der Aufbau der Materie*, edited by von K. W. Wagner, Friedr. Vieweg & Sohn, Braunschweig, 1949.

[205] Hückel, E.: *Quantentheoretische Beiträge zum Problem der aromatischen und ungesättigten Verbindungen. III.*, in: ZP 76 (1932), 628–654. I confine the following discussion to those aspects of this paper not considered in detail by Berson. Cf. Berson, J. A.: *Erich Hückel – Pionier der Organischen Quantenchemie* (1996).

[206] Hückel, E.: *Quantentheoretische Beiträge III.*, p. 628.

[207] Hückel provided a quantum-mechanical proof of this in the appendix to his paper. Cf. Ibid., pp. 643 f.

[208] Ibid., p. 637 (emphasis mine).

adequately described as mathematical abstractions in configurational space. This was precisely where his considerations on benzene had led him, underscoring once again its prototypical role among the class of aromatic compounds. In the previously mentioned letter to Born, Hückel emphasized this view:

> I have meanwhile calculated out naphthaline and anthracene as well using Bloch's method and arrive at entirely commensurate results. There is nothing in quantum theory corresponding to the formulas from the valence line diagram [...]. On the contrary, in configurational space in general, atoms lying symmetrically with reference to the mirror-image planes of the molecules are equivalent. Experiment demands this as well.[209]

The equivalence of the carbon atoms of fused ring compounds as confirmed by their chemical behavior was hence, Hückel argued, only properly described in abstract configurational space. Here again, the symmetry of the molecule under consideration was an essential starting point.

These considerations reveal Hückel's attempt to reduce the traditional conceptions of organic chemistry, marked by visual images of aromatic compounds, to profounder principles. He pointed to fundamental regularities and the logical structures underlying the representational forms used by organic chemists. Once again the influence of Hilbert's conceptions of science shines through.

2.2.4 The Quantum-Theoretical Basis of the Chemical Behavior of Substituted Benzene Derivatives

The breakthrough in the benzene problem had important repercussions on Hückel's further research practice. It was an effective stimulus for him to work on related topics. A few months after his first paper on the electron configuration of benzene and still before applying his method to other aromatic derivatives (as we saw in the foregoing section), Hückel submitted another paper to the *Zeitschrift der Physik* in which he attempted to explain the chemical behavior of monosubstituted benzenes according to quantum theory.[210] In his first paper he had explicitly mentioned this issue, emphasizing that his quantum-theoretical model would not be completely satisfactory unless it could explain the chemical behavior of polysubstituted benzene.[211]

Experiments had shown that a substituent at one location directed a second substituent into either the ortho, para or meta positions, depending on its specific character (Fig. 2.13).

---

[209] SBPK, Papers of Hückel, Box 6, Folder 5.21, Letter Hückel to Born, Cannstatt (Stuttgart) 29. November 1931.

[210] Hückel, E.: Quantentheoretische Beiträge zum Benzolproblem. II. Quantentheorie der induzierten Polaritäten, in: ZP 72 (1931), 310–337.

[211] Hückel, E.: Quantentheoretische Beiträge zum Benzolproblem. I. Die Elektronenkonfiguration des Benzols und verwandter Verbindungen, pp. 213–214.

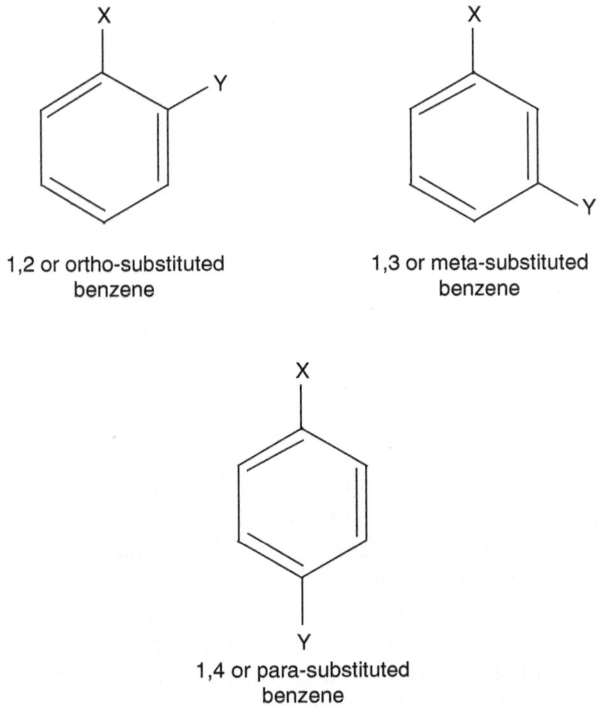

Fig. 2.13

Specifically, the various potential products of the reaction do not form in the same quantities and the rate of the reactions of formation depended on the temperature. For example, it was known that no measurable amounts of the meta product resulted from the nitration of chlorobenzene C_6H_5–Cl. Only ortho and para products were formed. In the nitration of toluene, C_6H_5–CH_3, although a low proportion of the meta products results, ortho and para-nitrotoluenes are the main products (Fig. 2.14)
.

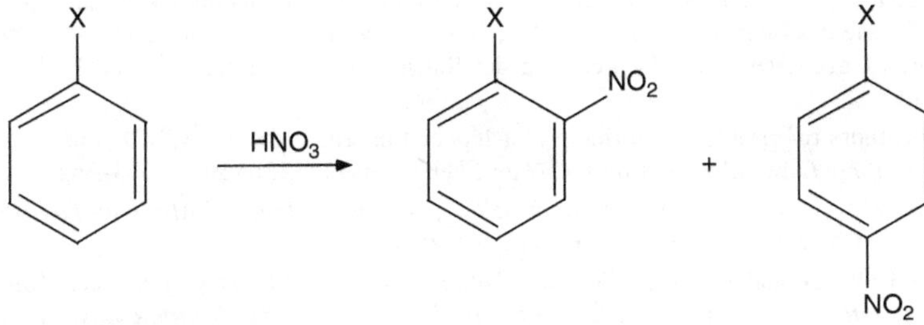

Fig. 2.14 X=Cl, CH_3

Fig. 2.15 X=NO_2, COOH

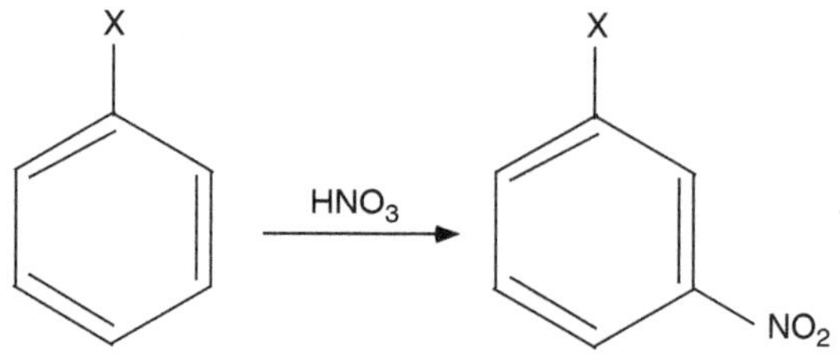

Organic chemists consequently spoke of the substituents –Cl and –CH_3 "directing" the ortho or para positions. Other substituents, such as, NO_2, and COOH direct the second substitution into the meta position. This means that the meta compound primarily forms, for instance, in the nitration or halogenation, etc., of nitrobenzene C_6H_5–NO_2 (Fig. 2.15) .

Various chemists tried to interpret the generally valid rules according to the electronic, electrostatic or polar theories of valence.[212] In his paper Hückel only mentioned the "theory of electrical contrasts" by Daniel Vorländer (1867–1941)[213] and the "theory of induced alternate polarities"[214] by Arthur Lapworth (1872–1941),[215] William O. Kermack (1898–1970) and Robert Robinson (1886–1975).[216]

Detailed discussion of the slightly differing conceptions of these authors is beyond the scope of this work.[217] But Hückel limited himself, in principle, to Vorländer's interpretation and its quantum-theoretical treatment. Hückel justified this preference for Vorländer's theoretical considerations as follows:

> Let us confine ourselves in the following to a discussion of Vorländer's interpretation, because the interpretations attempted by the other authors mentioned use a series of auxiliary concepts and hypotheses with meanings and justifications that are not always

---

[212]Cf. Brock, W. H.: *The Northon History of Chemistry*. W. W. Norton & Company, New York, 1993, Chap. 13 The Nature of the Chemical Bond, pp. 462–505 und Chap. 14 Structure and Mechanism in Organic Chemistry, pp. 506–569; Nye, M. J.: *From Chemical Philosophy to Theoretical Chemistry*. University of California Press, London, 1993; Stranges, A. N.: *Electrons and Valence: Development of the Theory, 1900–1925*. Texas A&M University Press, 1982, Chap. 6 The Decline of the Electrostatic or Polar Theory of Valence, pp. 164–200.

[213]Vorländer, D., Siebert, E. und Spreckels, E.: Die Lehre von den innermolekularen Gegensätzen und die Theorie des Benzols, in: B. 52 (1919), 263–283; Vorländer, D.: Die Lehre von den innermolekularen Gegensätzen und die Lenkung der Substituenten im Benzol II., in: B. 58 (1925), 1893–1914.

[214]Other authors referred to "the principle of induced alternating polarity." Cf. van Duin C. F.: *Über die künftige Entwicklung der organischen Chemie*, in: ZPC 130 (1927), 353–364.

[215]Lapworth, A.: *A Theoretical Derivation of the principle of Induced Alternate Polarities*, in: Journal of the Chemical Society (London) 121 (1922), 416–427.

[216]Kermack, W. O. and Robinson, R.: *An Explanation of the Property of Induced Polarity of Atoms and an Interpretation of the Theory of Partial Valencies on an Electronic Basis*, in: Journal Chemical Society (London) 121 (1922), 427–440.

[217]Cf. Nye, M. J.: From Chemical Philosophy to Theoretical Chemistry (1993).

immediately apparent, while Vorländer's is founded on a very simple scheme that does full justice to simple cases.[218]

Vorländer applied the old electrochemical theory of electrical contrasts suggested by Jöns Jakob Berzelius (1779–1848) in an effort to explain the substitution reactions of substituted benzenes. He imagined that upon introduction of a given substituent into the benzene ring its specific positive or negative electrochemical character induced a contrasting electric charge on the carbon atom to which it attached itself. This induction is then passed on to the other carbon atoms along the ring with the effect that neighboring carbon atoms are always of opposite charge. Vorländer postulated that the various organic substitution reactions for substituted benzenes are due to the magnitude of the electrochemical contrasts between the electropositive hydrogen atoms and the negatively induced carbon atoms forming the benzene ring. He indicated this intramolecular tension by adding + and – signs to the diagram.[219] Vorländer also tried to depict the relative magnitudes of the electrical contrasts of a given bond by the length of its valence lines: long valence lines signified a larger intramolecular tension than short ones depicting a lesser electrochemical contrast. The rate at which a second substituent reacted depended on the magnitude of the electrochemical contrast. The greater the contrast, the more rapidly the second substitution took place. Chlorobenzene and nitrobenzene are two examples (Fig. 2.16).

Accordingly, a secondary substitution of chlorobenzene was supposed to generate ortho-para positions because the carbon atoms at these positions were negatively induced by electronegative chlorine, consequently raising the intramolecular tension between the C and H atoms. Nitrobenzene, on the other hand, induced a meta substitution. Vorländer thought a greater electrochemical contrast affected the binding power of the H atoms, thus easing decomposition by the H atoms at ortho-para

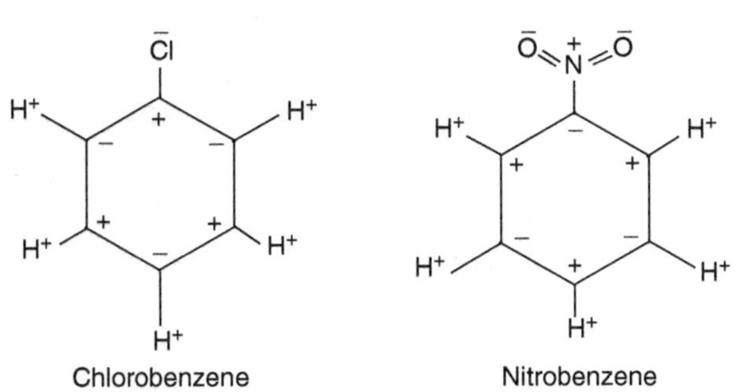

Fig. 2.16

[218]Hückel, E.: Quantentheoretische Beiträge zum Benzolproblem. II. Quantentheorie der induzierten Polaritäten, in: ZP 72 (1931), 310–337, p. 311.

[219]Cf. Vorländer, D.: *Säure, Salz, Ester und Addukt*, in: B 58 (1925), 118–143.

positions, in the case of chlorobenzene, or at a meta position, in the case of nitrobenzene. As a consequence there should be a strong tendency for substitution reactions. But Hückel detected a problem with this conclusion. As a result of the electrostatic effect between a negatively induced C atom and the positive H atom, one should expect a strengthening of the bond rather than its relaxation. Hückel specified in a footnote: "We do indeed see [from the quantum-theoretical interpretation] that Vorländer's interpretation should be modified on this point, to the effect that the charges 'induced' by the Cl atom on the C atoms take on precisely the opposite signs to Vorländer's."[220] For this reason Hückel preferred to refer to a theory of "induced polarities" instead of "electrical contrasts."[221] The title of his paper was specifically chosen with this in mind: "Quantum theory of induced polarities."

For his quantum-theoretical considerations Hückel retained Vorländer's idea that the introduction of a substituent to the benzene ring altered the molecule's overall charge distribution in a particular way. Because the substituted benzenes do not lose their aromatic character, Hückel regarded the effect induced by the substituents as a distortion, affecting primarily the charge distribution of the 6 $[p]_h$ electrons, because, in his opinion, they were the most loosely bound.[222]

Hückel's calculations according to quantum theory demonstrated that the distortion (change) in the distribution of charges of the 6 $[p]_h$ electrons resulted in precisely the opposite signs for each C atom on the benzene ring than prescribed either by Vorländer or by Lapworth and Robinson.[223] He illustrated his results with the following "images" (Fig. 2.17).

To calculate the influence of this change on possible further substitution reactions, Hückel made the following basic assumption: A reduction in the density of "induced" charge for $[p]_h$ electrons, i.e., a rise in the positive charge of a carbon

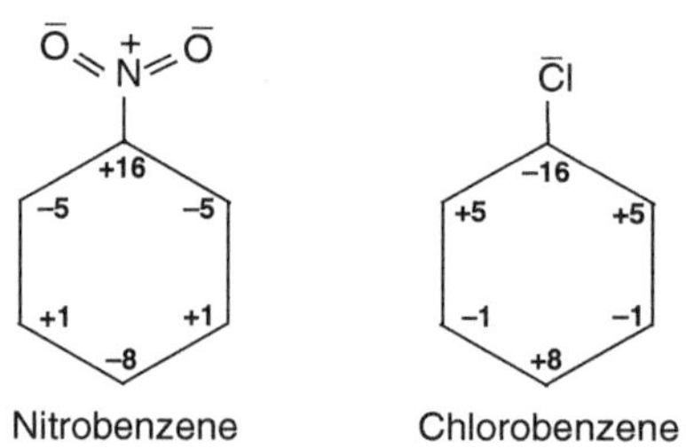

Fig. 2.17

[220]Hückel, E.: Quantentheoretische Beiträge zum Benzolproblem. II. Quantentheorie der induzierten Polaritäten, in: ZP 72 (1931), 310–337, p. 313, footnote 1.

[221]Ibid.

[222]Ibid., p. 315.

[223]Ibid., p. 324.

atom on the benzene ring, facilitates subsequent substitution, whereas an increase in the density of "induced" charge inhibits substitution.[224]

For example, in chlorobenzene there is a surplus of positive charge in the ortho-para carbon atoms as compared to benzene (cf. Fig. 2.17). Hückel contended that, owing to the Coulomb forces between the induced surplus charges and the positive H nucleus, a repulsive potential forms that in the case of chlorobenzene is enough to provide the energy needed to dislodge the H atom. Thus, the separation work of the H atom in the ortho-para positions is lowered and the equilibrium distance between the C and H atoms is increased. We do not need to go into the specifics of the reaction mechanism to get the gist of how, according to Hückel, the introduction of a second substituent at the ortho-para positions would be facilited.[225]

When Hückel's paper appeared, his assumption that an increase in negative charge would inhibit substitutibility immediately encountered strong opposition by Lapworth and Robinson. They published a joint communication in *Nature* on February 20, 1932 in which they put forward chemical reasons refuting Hückel's suggested mechanism and supporting their own diametrically opposite assumption that under those conditions substitutions were rather facilitated.[226] A few months later Hückel and his brother Walter sent their reply to the English authors in the form of a letter to the editor of *Nature*.[227] Compromise was out of the question for the Hückel brothers. They concluded with a reiteration of the main purpose of Erich's original paper:

> To give such a complete theory was not the intention of Hückel's paper. Its purpose was rather merely to show that for the influence of different groups on simple substitution reactions in benzene a simple and intuitive interpretation without detailed assumptions about the mechanism of reaction can be given by means of the quantum theoretical calculation of the distortion of the electronic distribution in the benzene ring produced by the substituent.[228]

The English authors were yet to be convinced. Their rebuttal came in the form of another letter to the editor of *Nature*.[229]

Meanwhile, in England Christopher Ingold was working on a general electronic theory of organic reactions.[230] Following Ingold, three effects were involved in the distortion in the distribution of charge on the electrons corresponding to Hückel's 6

---

[224]Ibid., p. 325 f. Cf. Hückel, E.: *Kritische Betrachtungen zur Theorie der Substitutionsreaktionen an substituierten Benzolen*, in: ZPC (B) 35 (1937), 163–192, p. 164.

[225]For more details on Hückel's explanation of the substitution reaction for monosubstituted benzene, see Ibid.

[226]Lapworth, A., Robinson, R.: *Distribution of Electrons in the Aromatic Nucleus and the Early Stages of Aromatic Substitutions*, in: Nature 129 (1932), 278. See Nye, M. J.: *From Chemical Philosophy to Theoretical Chemistry* (1993), p. 193.

[227]Hückel, E., Hückel, W.: *Theory of Induced Polarities in Benzene*, in: Nature 129 (1932), 937–938.

[228]Ibid., p. 938.

[229]Lapworth, A., Robinson, R.: *Theory of Induced Polarities in Benzene*, in: Nature 130 (1932), 273.

[230]Cf. Nye, M. J.: *From Chemical Philosophy to Theoretical Chemistry* (1983), pp. 163–211.

[p]h electrons: inductive, mesomer and polarizability effects.[231] A few years later, George W. Wheland and Linus Pauling took a fresh look at these effects pin-pointed by Ingold. Extending and modifying Hückel's theses, they attempted to construct a new theory based on quantum mechanics.[232] This sparked a controversy between Hückel, Pauling and Wheland that will be the subject of Chapters 3 and 4 below.

2.2.5 First Reactions to Hückel's Papers on Aromatic Compounds

The reaction of Hückel's mentor Debye to the first paper on the benzene problem was surprisingly negative. "When he heard about how long it was," Hückel recollected, "his response was: If the paper is that long, it can't possibly be right."[233] It was 83 pages long. Understandably, Hückel was "pained" by this comment.[234] But his expositional style had been molded by the evolving quantum-mechanical formalisms and the conventional theoretical approaches of chemists.[235] After carefully reading through a draft of the paper, Hund had sent Hückel many critical comments to improve his manuscript: "Couldn't you make your paper more readable by making your assumptions for the calculations more precise? Currently the reader is perhaps first led to the impression that he has understood a section, only to realize 10 pages later that he hasn't understood a thing. And couldn't it be shorter?"[236] Born also advised Hückel to slim down the calculation "according to the scheme used by Heitler and London" and to make it structurally more transparent.[237] He also suggested that the calculations "according to Bloch's procedure" be "simplified in a corresponding manner."[238]

Hückel doubtlessly implemented Hund's and Born's kind suggestions in the final versions of his papers. Yet his cumbersome style remained particularly opaque to chemists in the 1930s not yet conversant in the terminology of group theory

---

[231] Ingold, C. K. *Principles of an Electronic Theory of Organic Reactions*, in: Chemical Reviews 15 (1934), 225–274.

[232] Wheland, G. W. and Pauling, L.: *A Quantum Mechanical Discussion of Orientation of Substituents in Aromatic Molecules*, in: JACS 57 (1935), 2086–2095.

[233] Hückel, E.: *Erinnerungen an Peter Debye und an meine Lehrjahre*, in: Physikalische Blätter 28 (1972), 53–57, p. 55. Hertha Sponer confirmed this remark by Debye in an interview with T. S. Kuhn and M. Mayer. She later reminisced: Now (Hückel's) famous paper, (...) I know the following conversation between Hückel and Debye. Hückel said to Debye, "That paper becomes awfully long." You know it was a terrifically long paper, (...). And he told Debye about it, and Debye said, "Hückel, if you cannot say anything in a fairly short fashion, and need so many pages, it cannot be right." AHQP, Interview with J. Franck and H. Sponer, July 12, 1962, p. 14.

[234] Ibid.

[235] For the dissemination of new ideas in science, clarity and succinctness are useful tools. Hückel's style was evidently not particularly well suited for this purpose.

[236] SBPK, Papers of Hückel, Box 6, Folder 5.17, Letter Hund to Hückel, Leipzig, 8. Mai 1931.

[237] SBPK, Papers of Hückel, Box 6, Folder 5.11, Letter Born to Hückel, Göttingen 5. Dezember 1931.

[238] Ibid.

employed by physicists and physical chemists.[239] His heavy use of mathematical formalisms and failure to separate the essentials out from among less important details set additional hurdles for his readers. This was not without negative consequences for the dissemination of Hückel's ideas, particularly among chemists.

Nevertheless, soon after the appearance of his first paper on the benzene problem Hückel's new ideas were acknowledged in an article in the *Zeitschrift für angewandte Chemie* entitled "On organic double bonds" (received on August 6, 1931).[240] The author happened to be a classical organic chemist and proponent of modern atomic theory, Arthur von Weinberg (1860–1943).[241]

Weinberg's "historical sketch" offered his readers a survey of the advances in research he had witnessed since 1877 on the nature of double bonding in organic compounds until the publication of Hückel's "interesting paper [...] on the quantum-theoretical treatment of benzene compounds."[242] Although he did not go into any detail, the mere fact that this organic chemist mentioned Hückel's accomplishments

---

[239]Hückel shared this stylistic weakness with Mulliken, whose profound insights were embedded in a fine web of extremely tightly packed correlations and as a result were apparent only to the specially initiated. Cf. Mulliken, R. S.: *Bonding Power of Electrons and Theory of Valence*, in: CR, 9 (1931), 347–388; *Electronic Structures of Polyatomic Molecules and Valence. III. Quantum Theory of the Double Bond*, in: Physical Review 41 (1932), 751–758; *Electronic Structures of Polyatomic Molecules and Valence. IV. Electronic States, Quantum Theory of the Double Bond*, in: Physical Review 43 (1933), 279–302.

[240]von Weinberg, A.: *Über organische Doppelbindungen*, in: Zeitschrift für angewandte Chemie 44 (1931), pp. 815–818.

[241]Arthur von Weinberg was born on August 11, 1860 in Frankfurt am Main. His father Bernhard Weinberg was co-owner of the dye wholesaling business Leopold Cassella. Arthur commenced his scientific studies, notably in chemistry, physics and mathematics at Strasbourg in 1878. After completing his third semester he became assistant to Rudolf Fittig (1835–1910), the discoverer of pinacolone and diphenyl. 1880 he moved to Munich, where he concentrated his efforts on chemistry, becoming a pupil of Adolf von Baeyer (1835–1917). In July 1882 he defended his dissertation "On carbostyril." He continued on as A. von Baeyer's assistant for another quarter year before joining his father's firm in October 1883, where he synthesized various dyes until 1893. In 1904 he was one of the cofounders of "little IG" dye concern, Hoechst-Casella-Kalle. He continued to be influential in the formation of the dye trust, I.G. Farbenindustrie AG, becoming a member of the supervisory and managerial board. His Jewish origins compelled him to leave the managerial board in 1935 and two years later the supervisory board as well, when the National Socialist Workers Party was conducting its economic Aryanization campaign in Germany. On June 2, 1942, at 82 years of age, von Weinberg was arrested on the orders of the Bavarian area commander and sent to the concentration camp in Theresienstadt, where he died on March 3, 1943 after a gall-bladder operation. For further biographical details, cf. Gans, L. und Walden, P.: *Zum 70. Geburtstag von A. v. Weinberg*, in: Zeitschrift für angewandte Chemie 43 (1930), 703-708; Ritter, H. und Zerweck, W.: *Arthur von Weinberg, 1860-1943*, in: CB 89 (1956), XIX–XLI; *Weinberg, Arthur von*, in: *Lexikon bedeutender Chemiker*, Verlag Harri Deutsch, Thun 1989; Heine, J. U.: *Verstand & Schicksal – Die Männer der I.G. Farbenindustrie AG in 161 Kurzbiographien*, Weinheim 1990, p. 255 f.

[242]von Weinberg, A.: *Über organische Doppelbindungen*, p. 818. Weinberg was completely familiar with the benzene problem, having himself published four papers on the subject between 1919 and 1921 in the *Berichten der Deutschen Chemischen Gesellschaft*. Cf. Weinberg, A.: *Zum Benzol-Problem*, in: B. 52 (1919), 928–940; *Zum Benzol-Problem (II)*, in: B. 52 (1919), 1501–1508; *Zum Benzol-Problem (III)*, in: B. 53 (1920), 1353–1358; *Zum Benzol-Problem (IV):*

is of particular interest within our context. He even went so far as to say: "the future of scientific research on all kinds of bonds" also lies along the path opened by Hückel in quantum theory.[243] Weinberg concluded his piece with the far-sighted observation: "The future calling of physics will be to breathe life into our frozen chemical images."[244]

About a year after receiving von Weinberg's essay, the renamed journal *Angewandte Chemie*[245] published an article by Walter Hückel on "Advances in organic chemistry 1929–1931."[246] In his report Walter Hückel focused on the progress made in the individual fields of organic chemistry.[247] In the last part of his report, under the heading "Application of quantum mechanics to problems in organic chemistry," he also referred briefly to his brother's papers on the benzene problem. He emphasized the innovative finding that "the means of rendering double bonds" in conventional depictions of benzene "were no longer legitimate."[248]

These cursory references to Erich Hückel's extremely abstract papers obviously were not able to awaken the interest of chemists in Germany, whose training was generally geared toward experimental practice. But they nevertheless document a certain curiosity about the new orientation in chemical research.

In 1933 the second edition of the 24th volume in two parts of the *Handbuch der Physik* appeared.[249] Its blue cover earned it the nickname "the atomic physicist's blue bible."[250] The first part was devoted to *quantum theory* and included a survey article by Hund with the title "General quantum mechanics of atomic and molecular structure"[251] in which Hückel's quantum theoretical accomplishments concerning aromatic compounds also found their place.[252] Owing to the high proportion of theory and mathematics, Hund's valuable contribution was undoubtedly hard for practice-oriented organic chemists to digest.

Die Naphthalinformel, in: B. 54 (1921), 2168–2171; *Zum Benzol-Problem V: Der Benzolring in Substitutionsprodukte*, in: B. 54 (1921), 2171–2175.

[243] Von Weinberg, A.: *Über organische Doppelbindungen*, p. 818.

[244] Ibid.

[245] The journal *Zeitschrift für Angewandte Chemie* was renamed *Angewandte Chemie* in 1932.

[246] Hückel, W.: Die *Fortschritte in der organischen Chemie 1929–1931*, in: Angewandte Chemie 45 (1932), 457–471.

[247] His report was divided into the following parts: I. The limitations of conventional structural formulas. II. Spatial structure of organic molecules. III. Discovering the finer molecular structure by methods used in physics and physical chemistry. IV. On the nature of the chemical bond.

[248] Hückel, W.: Die Fortschritte in der organischen Chemie 1929–1931, p. 471.

[249] *Handbuch der Physik*, Band XXIV (zweiter Auflage), *Erster Teil: Quantentheorie, Zweiter Teil: Aufbau der zusammenhängenden Materie*. Verlag von Julius Springer, Berlin, 1933.

[250] Friedrich Hund, conversation with A. Karachalios, Göttingen, August 22, 1994. The first "bible" of atomic physicists was Arnold Sommerfeld's *Atombau und Spektrallinien*. Cf. Eckert, M.: *Die Atomphysiker*, pp. 59–60.

[251] *Handbuch der Physik* (zweite Auflage), Band XXIV, *Erster Teil: Quantentheorie*. Verlag von Julius Springer, Berlin, 1933, Chap. 4: Allgemeine Quantenmechanik des Atom– und Molekelbaues, pp. 561-693.

[252] Ibid., p. 690 f.

Besides being of interest to theoretical physicists, the contributions by Hückel are likely to have attracted physical and organic chemists working on theoretical problems in industrial laboratories. It was from some of these that the first positive reactions to his papers came in.

On May 25, 1933, at a meeting of the permanent board of the Deutsche Bunsen Gesellschaft for Applied Physical Chemistry, the topic of the 39th convention of the society, scheduled to meet in Bonn the following year, was under discussion.[253] The two most popular umbrella themes suggested were: "Textile chemistry" and "Application of physical chemistry to practical problems in organic chemistry." Georg Bredig, full professor of physical chemistry at the polytechnic in Karlsruhe was assigned the task of providing a list of relevant main topics and naming four or five referees. W. K. Friedrich Gaus, president of the supervisory board of I.G. Farben in Ludwigshafen, assumed the same task for the second proposal.[254]

At the board meeting that winter, on November 25, 1933, there was a fuller discussion of the convention topic. Gaus was unable to attend this meeting, so in his stead Otto Schmidt[255] delivered the report on the topic Gaus had chosen: "Valence,

---

[253] Present at this meeting were: Rudolph Schenck (president, see footnote 267), Max Bodenstein (professor of physical chemistry in Berlin), Friedrich Körber (treasurer), Friedrich Bergius (industrial chemist), Wilhelm Biltz (professor of inorganic chemistry at the polytechnic in Hannover), John Eggert (professor of photochemistry at the University of Berlin), Arnold Eucken (professor of physical chemistry at the University of Göttingen), W. K. Friedrich Gaus (president of the supervisory board of I.G. Farben), Dr. Theo Goldschmidt (industrial chemist), Georg Grube (professor of physical chemistry at the polytechnic in Stuttgart), Georg von Hevesy (professor of physical chemistry at the University of Freiburg), ? Pollitzer, Otto Schmidt (see footnote 255), Alfred Stock (professor of inorganic chemistry at the polytechnic in Karlsruhe), Arthur von Weinberg (see footnote 241), Wilhelm Bachmann (extraordinary professor of pure and applied colloidal chemistry at the polytechnic in Hannover). Others invited to attend were Andreas von Antropoff (professor of physical chemistry at the University of Bonn) and Georg Bredig (professor of physical chemistry at the polytechnic in Karlsruhe). BALM, Ordner DBG II, Vorstandsberichte 1933–1953, Bericht über die Sitzung des Ständigen Ausschusses der Deutschen Bunsen-Gesellschaft am 25. Mai 1933 zu Karlsruhe.

[254] BALM, Ordner DBG II, Vorstandsberichte 1933–1953, Bericht über die Sitzung des Ständigen Ausschusses der Deutschen Bunsen-Gesellschaft am 25. Mai 1933 zu Karlsruhe.

[255] Otto Schmidt was born on September 8, 1874 in Cologne. After attending a preparatory Gymnasium in the humanities in Cologne, he studied at the Universities of Bonn and Zurich from 1893 to 1897, taking his doctorate in Bonn in 1898. After a period as assistant in Zurich and later as an employee of a Swiss firm, he became assistant to Richard Anschütz in Bonn in 1901, where he worked toward his habilitation degree in organic and physical chemistry. On March 1, 1907, he accepted a position as chemist at BASF, where he rose from executive secretary in 1921 to director in 1925. After retiring in 1932 he continued to work at the main laboratory of I.G. Farbenindustrie AG in Ludwigshafen until 1938. There he conducted numerous quantum-theoretical analyses, in particular in the new field of quantum chemistry. He found the rule of double bonding, which he was able to interpret quantum-mechanically from the principle of spin distribution. He also introduced the box model into the theory of organic compounds. Later investigations treated the density distribution and zero-point energy of conducting electrons of graphite as well as of π electrons ("B electrons") of aromatic compounds. He died on May 17, 1943. See BALM, Ordner DBG II, Vorstandsberichte 1933-1953, Bericht über die Sitzung des Ständigen Ausschusses der Deutschen Bunsen-Gesellschaft am 25. Mai 1933 zu Karlsruhe.

structure and kinetics in the field of organic chemistry with special attention to technical processes."[256] The minutes of the board meeting document that Schmidt's report was favorably received. It emphasized the importance of the new quantum mechanics for the problem of valence in organic chemistry and the needs of the practical organic chemist:

> On the other hand, the problem of valence is something that the organic chemist definitely needs as a tool of research. In the dye laboratories of I.G. [Farben], the old Kekulé formula is still used as a basis for their entire research activities. One would like to have information on the labile electrons and the positions particularly susceptible to photochemical action. Furthermore, technicians would like to receive orientation on modern concepts of valence. There is such an abundance of noteworthy new conceptions in this area that organic chemists could very much profit by. [...]
>
> Mr. Schmidt summarizes briefly again that innovatively engaged organic chemists must gain new ideas from science. The basis of all their inventions is valence and modern quantum mechanics, particularly in the form developed by London and Heitler. Despite the partly quite concentrated mathematical treatment, it bears such a breadth of new aspects along with entirely lucid ideas that he believes it would be in the interest of organic chemists to become acquainted with this new field through an authority. Much more new and interesting things could certainly be conveyed, and Mr. Schmidt expresses the wish that this topic be made the guiding theme of the Bunsen Society and only the very best experts be enlisted to deliver the talks.[257]

In conclusion, Schmidt commented in detail on the agenda suggested by his company on "Problems of valence, structure and kinetics in the field of organic chemistry." 6 speakers were planned for the problem of valence, 10 for structural issues and 11 for kinetics. The proposed speakers on the problem of valence were Walther Kossel, Friedrich Hund, Hans Georg Grimm, Reinhard Mecke and Karl Lothar Wolf. Erich Hückel was the suggested speaker on the topic "the wave-mechanical nature of valence" in benzene, double bonding and radicals.[258]

During the discussion of Schmidt's motion, Andreas von Antropoff, full professor of physical chemistry at Bonn, informed those present that following a discussion with his fellow colleagues in inorganic chemistry he too wished to raise a similar motion.[259] Gustav Pfeiffer, extraordinary professor at Bonn for physiological chemistry, and Eduard Hertel, a temporary lecturer of physical chemistry at

---

[256]BALM, Ordner DBG II, Vorstandsberichte 1933–1953, Bericht über die Sitzung des Ständigen Ausschusses der Deutschen Bunsen-Gesellschaft am 25. November 1933 zu Berlin.

[257]Ibid.

[258]Ibid. Bericht mit Anlage: Leitthema der nächsten Hauptversammlung der Deutschen Bunsen-Gesellschaft. Probleme der Valenz, Struktur und Kinetik auf dem Gebiete der organischen Chemie.

[259]Since 1930 some inorganic chemists had started thinking that they ought to have some understanding of Schrödinger's equation. This was motivated by the fact that the magnetic properties of inorganic substances were of particular interest to them, and they could not hope to make any further progress without any knowledge about the physics. The inorganic chemists Wilhelm Biltz and Wilhelm Klemm, both teachers at the polytechnic in Hannover, were especially interested in the rare earths and their magnetic properties. With this special interest in mind they approached their colleagues at Hannover, the physicists Erwin Fues and Hans Hellmann, with the intention of learning about the new concepts in quantum theory and their usefulness in chemistry. They

Kiel, had worked out the specifics of the proposed topic: "Applications of physical chemistry to practical problems in organic chemistry." The following five subtopics were envisioned:

(1) the concept of valence
(2) affinity splitting
(3) problems in stereochemistry
(4) wave mechanics of organic compounds
(5) kinetics of photochemical processes.[260]

The suggested speakers were each first-rate leaders in their specialties. For example, Walter Hückel or the inorganic chemist Wilhelm Biltz were named for the concept of valence, Karl Freudenberg or Walter Hückel for stereochemical problems, Max Bodenstein, Arnold Eucken or Karl Friedrich Bonhoeffer for the kinetics of photochemical processes. It is noteable that Erich only comes into consideration for the topic "wave mechanics of organic compounds."

Enthusiasm about the two proposals was not the only response among the majority of those in attendance. Above all they were considered too theoretical and not in accordance with the wishes of the technically oriented membership. "Nor did they meet the demands of the day, which call for a correlation of science with practice."[261] In Bodenstein's opinion, they were "better disposed for a textbook on organic chemistry from the standpoint of physical chemistry" than for a guiding theme of a conference.[262] He thought more specialized topics were necessary that would relate the scientific theory with practice as just mentioned, from the standpoint of physical chemistry. Another leading physical chemist and pioneer of chemical physics in Germany, Eucken, was of the same opinion. He would rather not make the problem of valence the guiding theme of the conference because the theories on which it was based "were still too uncertain, making the possibility of a bridge to practice too slight."[263] From this realistic standpoint Eucken suggested a compromise solution: He suggested accepting the proposal by von Antropoff as an introductory talk "around which the other talks could then crystalize."[264]

---

and their coworkers attended a special lecture offered by Fues for two semesters with this purpose in mind. Despite Hellmann's extra effort to translate the difficult concepts of quantum theory into "the language of chemistry," Fues's lectures only gave a vague idea of the usefulness of quantum mechanics to chemistry. Werner Fischer, conversation with W. H. Eugen Schwarz and A. Karachalios, Freiburg im Breisgau, July 4, 1999.

[260] BALM, Ordner DBG II, Vorstandsberichte 1933–1953, Bericht über die Sitzung des Ständigen Ausschusses der Deutschen Bunsen-Gesellschaft am 25. November 1933 zu Berlin.

[261] Ibid.

[262] Ibid.

[263] Ibid.

[264] Ibid.

Mark, an invited guest at this meeting, mentioned "that it would undoubtedly be very important to meet the demands of the chemical industry."[265] To achieve this he suggested the following two subject areas:

(1) issues of hydrogenation and cracking
(2) energy transfer to the surface

Mark stressed in addition that such a theme would meet "the legitimate demands of industry." He was also of the opinion that the "introductory speaker should offer industry a survey of the measurement methods developed in science. This was more important than theoretical considerations."[266] After discussing Mark's suggestion over, the following guiding theme was "unanimously resolved":

(1) cracking and hydrogenation
(2) energy transfers to surfaces

At the next meeting just prior to the conference, Rudolf Schenck[267] as first president of the Bunsen Gesellschaft since Adolf Hitler's rise to power informed those in attendance that the chemical industry had reservations about the agreed upon guiding theme and "one would therefore have to arrive at another solution." He suggested the following main theme: "Tasks and goals of physico-chemical research in organic chemistry."[268] Von Antropoff reported that he had drafted a

[265] Ibid.

[266] Ibid.

[267] Rudolf Schenck was born in Halle on March 11, 1870. In 1889 he started his university studies in science there, later concentrating on chemistry. In 1894 he took his doctorate under Jacob Volhard in Halle and continued to work for him as assistant until 1897. In that year he qualified for academic teaching (Habilitation) at Marburg and two years later became department director at the local institute of physical chemistry. In 1906 he accepted the chair in this field at the polytechnic in Aachen. In 1910 he was transferred to the newly founded polytechnic in Breslau (now Wroclaw in Poland) and soon was appointed its founding president. From 1916 until his retirement in 1935, Schenck was professor of chemistry in Münster. From 1933 until 1935 he was the first president of the Bunsen Gesellschaft since Adolf Hitler's seizure of power. As emeritus he continued to work until 1950 at a state-run research institution for metallochemistry in Marburg that received funding from various firms and organizations.

Schenck's first scientific investigations involved liquid crystals. He was primarily known, however, for his work on heterogenous equilibria between metals and their ores, whereby he applied ideas from physical chemistry to metallurgy (blast-furnace processes). He died in Aachen on March 28, 1965. For further biographical details see Fricke, R.: *Rudolf Schenk zum 70. Geburtstag*, in: Z. Elektrochem. 46 (1940), 101–105; Meinel C.: *Die Chemie an der Universität Marburg seit Beginn des 19. Jahrhunderts: Ein Beitrag zu ihrer Entwicklung als Hochschulfach.* Academia Marburgensis, ed. von der Philipps-Universität Marburg, Bd. 3, Elwert, Marburg 1978, p. 250 and 362; *Schenck, Rudolf*, in: Lexikon bedeutender Chemiker, Verlag Harri Deutsch, Thun 1989; Jaenicke, W.: *100 Jahre BUNSEN-Gesellschaft, 1894–1994.* Steinkopff, Darmstadt 1994, pp. 179–180.

[268] BALM, Ordner DBG II, Vorstandsberichte 1933–1953, Bericht über die Sitzung des Ständigen Ausschusses der Deutschen Bunsen-Gesellschaft am 17. Mai 1934 zu Bonn a. Rh.

major plan around this theme, "the main parts of which was supposed to address the spatial conception of organic compounds, and on the other hand, the energetic issues. Messrs. Dimroth (Würzburg) and Hückel (Stuttgart), whom he had envisioned for this alternative, had declined, however." Consequently these topics had to be abandoned.[269]

Why Hückel declined the invitation to the XXXIXth general meeting of the Bunsen Gesellschaft that took place in Bonn from May 17 to 20, 1934, is unclear. But it is noteworthy that the physical chemist Eduard Hertel nonetheless briefly acknowledged Hückel's achievements. In what context was this mention made?

Before closing his speech for the second time with the threefold salute "Sieg Heil!",[270] he emphasized that physical chemistry "currently had to exert much effort to maintain the existing level of efficiency."[271] This referred to some "tangible vacancies in our laboratories"[272] that had formed during the course of 1934. According to Schenck, "it was the task and duty of the German Bunsen Society to

---

[269]Ibid.

[270]The following passages taken from Schenck's introductory speech give a taste of the president's political views: "When in the elapsed year we convened at our festive inaugural meeting in Karlsruhe, we greeted Adolf Hitler's Reich for the first time. Today he is a familiar figure to us; we know his strength and the effective prowess with which he tackled problems to imbue a divided nation, that in many quarters was close to desperation, faith in itself again and to open its eyes to the fact that all its members and all its stations form a community and a unique organism to which all are inextricably connected, their lives long, for better or for worse" [...]

The deeply rooted goal of German scientific societies, not least of the German Bunsen Society for Applied Physical Chemistry, is an altruistic development of new forces for the common weal, thereby facilitating pinnacle achievements. The German nation can rest assured that, in future too, it will throw itself fully onto its task, applying its own means to help build a happy future. Thus it joins the mighty ranks of laborers in confronting the problems of our times, including research over the long term, whose fruits shall ripen only at a later time. [...]

It is with pride that the Bunsen Society can point to the valuable things German physical chemistry has conjured up and fabricated, and that its members have contributed, in peace and in war, for the protection of our land, for the fertilization of our fields, the exploitation of our energy resources and the refinement of our raw materials. [...]

The National Socialist Reich safeguards what is firmly grounded and deeply rooted. Thus we may rest in the firm conviction that its powerful protection will also fall to the share of science, an existential necessity for the institutions of the state and the nation, taking duly into account the natural – physical and intellectual – conditions, which secure the progress and primacy of science for the future as well. [...]

In this confidence I welcome the authorities of the Reich and of the state, who have given us the honor of their visit, as well as the Rhenish leaders of the party and its organizations, who opened the way for the Führer and, by the grace of God, helped him grasp the helm of the Reich." Schenk, R.: *Begrüßungsansprache zur XXXIX. Hauptversammlung der Deutschen Bunsen-Gesellschaft für angewandte physikalische Chemie*, in: Z. Elektrochem. 40 (1934), 401–404. p. 402. On the relations between the Bunsen Gesellschaft and National Socialism and its internal workings during this period, see Jaenicke, W.: *100 Jahre BUNSEN-Gesellschaft, 1894-1994*. Steinkopff, Darmstadt 1994.

[271]Ibid.

[272]Ibid.

act and to counter the loss by activating our remaining staff."[273] Hertel expanded on this idea in his main talk bearing the title: "Physico-chemical problems in organic chemistry."[274]

Right at the beginning of his talk, Hertel underscored "new impulses" that research on organic issues in physical chemistry had received: from evaluations of the interference of x rays, electrons and protons, the measurement of dipole moments, analysis of absorption spectra, a study of the Raman effect, magnetic susceptibility, quantum theory and quantum mechanics.[275] In his opinion, the success was "not always 100%," however. The reasons lay in a certain alienation between physics and organic chemistry. This problem of communication between physicists and organic chemists appeared to Hertel most severe in Germany. Traditional training of chemists, particularly into the 1920s, offered few points in common for contact with physicists. Hertel observed further that physicists often regard the "gigantic theoretical structures" of organic chemistry with skepticism and had a tendency to overvalue individual measurement results.[276] The organic chemist, on the other hand, "easily falls into the trap of performing physical measurements himself that appear useful, without respecting the difficulties posed in obtaining reliable results."[277] Hertel thus went beyond pointing out the importance of physical measurement to argue the necessity for an interdisciplinary approach between organic chemistry and physics in order to overcome the problems facing organic chemistry: "Whoever wants to solve problems in organic chemistry using the methods of physics must be at home in the conceptual world of organic chemistry and be properly schooled in the physical methods."[278]

Without a doubt, the first condition was a commonly held attitude among organic chemists for solving problems in physico-chemical research. But that was not enough. Organic chemists now needed training in modern mathematics and physics in order to be able to understand the new descriptions of binding states by quantum theory (such as for the aromatic compounds).

Altogether, Hertel considered it practical to distinguish between an "inner" and "outer" field in describing the physico-chemical behavior of a molecule. All effects taking place within a molecule among the atoms (intramolecular forces) belonged within the inner field. The effects working from the molecules outwards (intermolecular forces) are ascribable to the outer field. Hertel's explication of this was where his mention of Hückel's quantum-theoretical achievements occurs:

---

273 Ibid.

274 Hertel, E.: *Physikalisch-chemische Probleme der organischen Chemie*, in: Z. Elektrochem. 40 (1934), 405–413.

275 Ibid., p. 405.

276 Ibid.

277 Ibid.

278 Ibid.

> As concerns the theoretical treatment of inner molecular fields, it seems the new quantum mechanics, especially the line followed by Messrs. Erich Hückel and Linus Pauling, is destined to lead to a decisive solution to the problem.
>
> The organic chemist will always demand, however, chemical experiment as the touchstone for any theory. The usefulness of a theory is demonstrated to him when the theory can confirm his observations, and furthermore allows him to make predictions, in the ideal case, even to make advance calculations.[279]

Although Hertel was thus passing favorable judgment on "the new quantum mechanics," experiment basically remained for him the decisive element in chemical research.

Despite the initial positive reactions to Hückel's quantum-theoretical treatment of problems in chemistry discussed above, on the whole, after 1934 Germany's chemists showed little interest in this new approach to chemical issues. One important reason is clear. An experimental and heuristic approach predominated among German chemists.[280] Between the two World Wars, theory generally assumed a subordinate role in German chemistry. Organic chemists were particularly conservative in applying physical theories and methods to their field. This was largely due to their academic training.[281] This traditional schooling stood in the way of an ability to draw connections between familiar chemical conceptions and an unintuitive, purely mathematical approach to problems. Other reasons for this sluggish reception of Hückel's ideas can be found in institutional, political, economic and ideological factors prevailing during the Nazi regime.[282] Before delving into this topic, it seems appropriate to first examine the controversy between Hückel and Pauling.

---

[279] Ibid., p. 408.

[280] Frenking, G.: *100. Geburtstag von Erich Hückel*, in: Chemie in unser Zeit 31 (1977), 27-31; Berson, J. A.: *Erich Hückel – Pionier der Organischen Quantenchemie*, Kap. 6. Die frostige Aufnahme durch die Praktiker.

[281] Johnson, J. A.: *Academic – Industrial Relations and Chemical Education in Germany, 1919–1939* (draft paper presented to Fourth International Conference of the CHMC, "Industrial – Academic Relationships in the Chemical and Molecular Sciences", October 4, 2002). I am grateful to Prof. Jeffrey Allan Johnson for kindly sending me a copy of his manuscript.

[282] See Chapter 5.

Chapter 3
The Controversy Between Erich Hückel and Linus Pauling over the Benzene Problem

Shortly after Hückel's quantum-theoretical work on the problem of aromatic compounds was published, the first paper on the same subject by the American Linus Pauling also appeared.[1] It was the fifth installment of a total of seven that Pauling published between April 1931 and July 1933 under the general title *The Nature of the Chemical Bond*.[2] This fifth part was the first coauthored with his pupil George Wheland, a National Research Fellow in Pasadena. In their quantum mechanical treatment of benzene, naphthaline and free organic radicals, they applied a "VB" approximation slightly different from Hückel's "first method" yet sharing some resemblance with it. The two subsequent parts of Pauling's series were copublished with John Sherman. These authors analysed thermochemical data to calculate the resonance energies of a large number of organic molecules, aromatic and conjugated systems.

In the fifth part, Pauling and Wheland presented the methodological difference between their approximative procedure in their quantum mechanical analysis of benzene and Hückel's as follows:

> E. Hückel has made a valuable start in this direction in a series of papers on the quantum mechanics of benzene. His method of attack, however, is very cumbersome. In this paper we present a treatment of the problem which is rather closely similar to that of Hückel's and which leads to the same result in the case of benzene, but in which the calculations are simplified to such an extent that the method can be extended to the naphthalene

---

[1]Pauling, L., Wheland, G.: *The Nature of the Chemical Bond. V. The Quantum Mechanical Calculation of the Resonance Energy of Benzene and Naphthalene and the Hydrocarbon Free Radicals*, in: JCP 1 (1933), 362–374.

[2]Pauling, L.: *The Nature of the Chemical Bond. Application of Results Obtained from the Quantum Mechanics and from a Theory of Paramagnetic Susceptibility to the Structure of Molecules*, in: JACS 53 (1931), 1367–1400; *The Nature of the Chemical Bond. II. The One – Electron Bond and the Three – Electron Bond*, in: JACS 53 (1931), 3225–3237; *The Nature of the Chemical Bond. III. The Transition from one Extreme Bond Type to Another*, in: JACS 54 (1932), 988–1003; *The Nature of the Chemical Bond. IV. The Energy of Single Bonds and the Relative Electronegativity of Atoms*, in: JACS 54 (1932), 3570–3582; Pauling, L., Sherman, J.: *The Nature of the Chemical Bond. VI. The Calculation from Thermochemical Data of the Energy of Resonance of Molecules Among Several Electronic Structures*, in: JCP 1 (1933), 606–617; *The Nature of the Chemical Bond. VII. The Calculation of Resonance Energy in Conjugated Systems*, in: JCP1 (1933), 679–686.

A. Karachalios, *Erich Hückel (1896–1980)*, Boston Studies in the Philosophy of Science 283, DOI 10.1007/978-90-481-3560-8_3,

molecule without undue labor. Furthermore, with the aid of additional simplifications and approximations we have been able to treat the problem of free radicals and to obtain results in surprisingly good qualitative agreement with experiment.[3]

Hückel himself later conceded that Pauling and Wheland replaced his quite ungainly first approximation (his "first method") with a "more elegant mathematical method"[4] that originated from the physicists Heitler and Georg Rumer and was subsequently developed by the mathematician Hermann Weyl at Göttingen. Hückel also agreed that this method allowed a "simpler and more straightforward" first approximation of the correlations.[5] As Pauling and Wheland pointed out, it involved a mathematical simplification of Slater's procedure for resolving the secular problem: "The benzene molecule can now be treated very simply by the Slater method, with the help of the rules formulated by one of us [Pauling] for finding the matrix elements occurring in the secular equation."[6]

Pauling had laid down these rules in a separate article that appeared a few months previously in the first volume of the *Journal of Chemical Physics*.[7] He combined a method for solving the secular equations, or secular determinants, with his own interpretation of the underlying mathematical formalisms of spin-invariant theory that Weyl and a few physicists had developed to correlate the valence-line diagrams used in chemistry.[8]

Pauling's physical interpretation of spin-invariant theory, i.e., his quantum-mechanical theory of resonance, as presented in this fifth part by Pauling and Wheland, triggered a controversy with Hückel that will be the subject of this chapter. We start with a brief historical sketch of the emergence of spin-invariant theory and its various physical interpretations. This will give us insight into the historical context in which the controversy arose and set the stage for Pauling's dominant role in it. The precarious position that quantum chemistry held in the German scientific community of the 1930s also emerges. A comparison of the divergent physical interpretations of the underlying mathematical formalisms by the two proponents also provides a fuller picture of the Hückel vs. Pauling controversy over the quantum-mechanical theory of resonance in chemistry. Finally, the tactics Hückel, Pauling and their collaborators used also throws light on their differing conceptions of the role of theory in chemistry and the methodological status of empirical observations.

---

[3]Pauling, L., Wheland, G.: *The Nature of the Chemical Bond. V.*, p. 363.

[4]Hückel, E.: *Grundzüge der Theorie ungesättigter und aromatischen Verbindungen*, in: Zeitschr. Elektrochem. 43 (1937), 752–788, p. 759.

[5]Ibid.

[6]Pauling, L., Wheland, G.: *The Nature of the Chemical Bond. V.*, p. 364.

[7]Pauling, L.: *The Calculation of Matrix Elements for Lewis Electronic Structures of Molecules*, in: JCP 1 (1931), 280–283.

[8]At the end of his paper Pauling underscored the usefulness of the above-mentioned methods: "The methods developed in this paper have been applied in a discussion of the structure of aromatic substances, free radicals, etc., to be published soon..." Ibid., p. 283. In the subsequent literature, the approximative method Pauling used is referred to as the valence-bond (VB) or Heitler-London-Slater-Pauling (HLSP) method.

3.1 Introduction: Calculating Perturbations with Spin Invariants and the Valence-Line Diagram in Chemistry

As we have seen in the preceding chapter, in 1930 Slater developed a procedure for calculating the interactions between large numbers of atoms without resorting to group theory.[9] Shortly afterwards, on July 29, 1930 the *Zeitschrift für Physik* received a paper by Max Born in Göttingen entitled "On the quantum theory of chemical forces."[10] It applied Slater's method to derive the formula that Born's assistant Heitler[11] had obtained by means of group theory for the binding energy of diatomic molecules.[12] Born's purpose was to eliminate group theory from the quantum mechanical theory of valence, a formidable obstacle for physicists and chemists alike in studying the literature on quantum chemistry. His intention beyond disseminating Slater's ideas was to derive "very succinctly and at a very elementary level a few other results of quantum theory on chemical bonds that Heitler and London had laboriously coaxed out of group theory."[13]

Early the next year, Heitler and Rumer at Göttingen analysed a series of molecules including hydrogen cyanide, cyanic acid, cyanogen and hydrazine according to the methods devised by Slater and Born, and calculated their binding energies. Their results were published in a joint paper under the title "Quantum theory of the chemical bonding of polyatomic molecules" received by the editors of *Zeitschrift für Physik* on January 17, 1931.[14] Benefiting from the useful advice of the local mathematician Hermann Weyl, Heitler and Rumer were able to present a way to reduce Slater's perturbation matrices for the secular problem into irreducible components from which the binding energy could be calculated relatively simply.[15]

Inspired by Heitler's report on his and Rumer's results at a seminar in Göttingen, Weyl soon afterwards found a more fundamental connection between the calculational methods employed by Heitler and Rumer and general principles of the group

---

[9]See Section 2.2.1.

[10]Born, M.: *Zur Quantentheorie der chemischen Kräfte*, in: ZP 64 (1930), 729–740.

[11]Heitler, W.: *Zur Gruppentheorie der homöopolaren chemischen Bindung*, in: ZP 47 (1928), 835–858.

[12]About Heitler's interest about group theory and their applications to the problems of chemical valence see: Gavroglu, K.: *Fritz London: A Scientific Biography*. Cambridge University Press, Cambridge, 1995, Chap. 2, *Polyelectronic Molecules and the Application of the Group Theory to Problems of Chemical Valence*, pp. 53–57; Gavroglu, K., Simoes, A.: *The Americans, the Germans, and the Beginnings of Quantum Chemistry: The Confluence of Diverging Traditions*, in: Historical Studies in the Physical Sciences 25(1) (1994), 47–110, *Polyelectronic Molecules and Group Theory*, pp. 66–70.

[13]Born, M.: *Zur Quantentheorie der chemischen Kräfte*, p. 729.

[14]Heitler, W., Rumer, G.: *Quantentheorie der chemischen Bindung für mehratomige Moleküle*, in: ZP 68 (1931), 12–41. A preliminary version of the paper appeared under the title "*Quantenchemie mehratomiger Moleküle*" in: Nachrichten von der königlichen Gesellschaft der Wissenschaften zu Göttingen. Mathematisch-physikalische Klasse aus dem Jahre, 1930, pp. 277–284.

[15]Ibid., p. 24.

and invariant theories and how they related to the valence-line diagrams used in chemistry.

Weyl published his results in two parts under the general title "On the quantum-theoretical calculation of molecular binding energies." It appeared at the turn of the year 1930/31 among the mathematical and physical reports of the scientific society at Göttingen.[16] His profound mathematical arguments proved that so-called "binary invariants" were the appropriate tool for calculating the binary energies of polyatomic molecules along the lines of the Heitler-London approximation.

Heitler and London had already shown in their fundamental paper[17] that, according to the new quantum mechanics, attractive forces with qualities similar to saturation existed between like atoms. This saturation of valences between two atoms corresponds to the interaction between two valence electrons (one for each atom) of antiparallel spin. This basic idea was confirmed not only for pairs of atoms of similar type, such as the hydrogen molecule (H–H), but also for forming diatomic molecules with many saturated valences.[18] Up to this point everything seemed to fit quite well into the Heitler-London theory, but triatomic molecules posed considerable mathematical complications. The main problem was combinatorial because the number of valence-line diagrams in chemistry does not equal the number of all the possible spin configurations. Weyl's analyses elucidated this fundamental combinatorial problem.

The following will be an elementary outline of Weyl's basic idea. His first step involved replacing the chemist's valence-line diagram with mathematical notation. First, each atom (A, B, C, ...) of a molecule is assigned a letter x, y, z, ... Then, every valency effect between two atoms, i.e. (A–B), is assigned the form [x, y] instead of the connecting dash x–y; free valences are rendered [x, l] to symbolize a valency effect with empty space [*Leerem*]. The terms corresponding to the different bonds and free valences of a molecule are then assembled as a product. A specific chemical formula (valence-line diagram) thereby corresponds to a specific product of the letters mentioned above (bracketed symbols) and vice versa.

[16] Weyl, H.: *Zur quantentheoretischen Berechnung molekularer Bindungsenergien*, in: Nachrichten von der königlichen Gesellschaft der Wissenschaften zu Göttingen. Mathematisch-physikalische Klasse aus dem Jahre, 1930, pp. 285–294; *Zur quantentheoretischen Berechnung molekularer Bindungsenergien. II*, in: Nachrichten von der königlichen Gesellschaft der Wissenschaften zu Göttingen. Mathematisch-physikalische Klasse aus dem Jahre, 1931, pp. 33–39.

[17] Heitler, W., London, F.: *Wechselwirkung neutraler Atome und homöopolare Bindung nach der Quantenmechanik*, in: ZP 44 (1927), 455–472.

[18] London, F.: *Zur Quantentheorie der homöopolaren Valenzzahlen*, in: ZP 46 (1928), 455–477; *Zur Quantenmechanik der homöopolaren Valenzchemie*, in: ZP 50 (1928), 24–51. Heitler, W.: *Zur Gruppentheorie der homöopolaren chemischen Bindung*, in: ZP 47 (1928), 835–858. See also Heitler's summary: Heitler, W.: *Der gegenwärtige Stand der quantenmechanischen Theorie der homöopolaren Bindung*, in: PZ 31 (1930), 185–204. In these papers Heitler formulated an approximation for the binding energy between two arbitrary atoms that, as he emphasized, corresponded to the formation of an electron pair of two electrons belonging to different atoms, hence corresponding to the "chemical valence line": A series of states result in which one, two or more pairs of electrons become saturated until all of an atom's valence electrons are used up.

The following example will illustrate this. The chemical formula

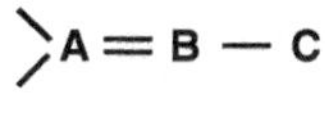

represents one double bond between A and B, one single bond between B and C, and the whole structure has in addition two free valences. Following Weyl, this formula can be replaced by the following symbolical product:

$$\varphi = [x,l]^2 \cdot [x,y]^2 \cdot [y,z]$$

The product φ of the bracketed symbols acquires real physical meaning when the spin characteristics of the bonding electrons are taken into account according to the Pauli principle. If you have two valence electrons, they can be assigned two spin vectors $x(x_+, x_-)$ and $y(y_+, y_-)$, where $x_+ = +1/2$, $x_- = -1/2$, $y_+ = +1/2$ and $y_- = -1/2$ are the corresponding components of the spin vector. According to the "first fundamental law of invariant theory," all physical properties based on the interaction between these two electronic spins can only depend on the "algebraic invariant," namely the determinant (spinor product):

$$[xy] = x_+y_- - x_-y_+$$

In general, every product of such bracketed symbols is called a monomial invariant. It is fully described by a diagram in which every vector x, y, . . . l is represented by one point and every bracketed symbol [x y] by one straight line connecting the points x and y. A bracketed symbol containing the free vector l [x l] can be represented instead by an open-ended line departing from x. Thus the monomial invariants correspond equivalently to chemical valence lines. Every invariant represents the state of spin of a molecule. That is why Weyl suggested calling possible energetic states of a molecule belonging to a spin invariant a "pure valence state." He wrote in this regard in his second paper:

> According to the *first fundamental law of invariant theory*, the only fundamental invariant from which all the others are quite logically constructed is the determinant dependent on the two vectors x, y:
>
> $$[xy] = x_+y_- - x_-y_+$$
>
> Each invariant that is completely rationally dependent on the vectors l, x, z, . . ., is therefore a linear combination of monomials, i.e., of products of the potentials of the fundamental invariants [l x],[l y],[x y], ... Such a monomial represents [. . .] one possible state of our molecule, which I denote as a *pure valence state*. By depicting each factor [x y] as a valence line between the atoms x and y, while representing a factor of the type [l x] by a

"free" valence line connecting atom x with "empty space," *the distribution of valence lines reflects in a very genuine way the pure valence state.*[19]

Accordingly, every valence-line diagram in chemistry corresponds to a "pure valence state," characterized by a specific spin invariant ϕ, and all possible combinations of the valence lines are identical with the combinatorics of the invariants of spin vectors. We thus basically have a formal analog for invariant theory to the valence-line diagram in chemistry.[20] Weyl's concept of "pure valence states" was a physical interpretation of the valence-line diagram in the formal mathematical language of monomial invariants. In principle, as we shall see more clearly below, Weyl developed a physical interpretation for the addition of invariants, indeed, for the formalisms of invariant theory as a whole. Weyl considered this theory an essential part of the quantum-mechanical theory of chemical bonding. But how did

---

[19] Weyl, H.: *Zur quantentheoretischen Berechnung molekularer Bindungsenergien. II*, in: Nachrichten von der königlichen Gesellschaft der Wissenschaften zu Göttingen. Mathematisch-physikalische Klasse aus dem Jahre (1931), 33–39, p. 36 (Original emphasis.)

[20] Much earlier, various 19th-century mathematicians had noticed this formal similarity between the valence-line diagram in chemistry and binary invariant theory. They realized that the combinatorics of valences used in the construction of chemical formulas, particularly in determining possible isomers, was the same as the combinatorial techniques in the mathematical area of invariant theory of binary forms. The first papers along these lines were written by the American Jean Joseph Sylvester and the English mathematician William Kingdon Clifford, who published the results of their research 1878 in the first volume of the journal Sylvester had founded, *American Journal of Mathematics Pure and Applied*, at the Johns Hopkins University in Baltimore. They were succeeded by the German theoretician of invariants Paul Gordan and the Russian chemist W. Alexejeff. Cf. Sylvester, J. J.: *On the Application of the New Atomic Theory to the Graphical Representation of the Invariants and Covariants of Binary Quantics*, in: American Journal of Mathematics 1 (1878), 64–90; Clifford, W. K.: *Extract of a Letter to Mr. Sylvester from Prof. Clifford of University College, London*, in: American Journal of Mathematics 1 (1878), 126–128; Gordan, P., Alexejew, W.: *Übereinstimmung der Formeln der Chemie und der Invariantentheorie*, in: ZPC 35 (1900), 610–633; Alexejeff, W.: *Über die Bedeutung der symbolischen Invariantentheorie für die Chemie*, in: ZPC 36 (1901), 741–743. Sylvester first noticed a similarity between invariant theory and the then still young atomistic theory of chemistry. Clifford elaborated on this insight and developed his own symbolical representation for various molecular compositions. Gordan and Alexejeff, for their part, expanded on the relation between the concepts and formulas used in chemistry and formulas of invariant theory. They showed that the formal methods of the symbolical theory of invariants were well suited to addressing the formal questions of modern chemistry and had close associations with the methods employed by atomistic structural theory in chemistry. Alexejeff's analyses offer this latter theory new tools for solving some formal questions, such as calculating the number of isomers and their structures. Chemists took little notice of these papers, however. Weyl provides the following reason: "Nevertheless, chemists stood by their familiar valence diagrams, for there was no physical interpretation for the addition of invariants and for dynamic laws by which the binding powers and the real stationary states could be determined. Today we can see that only such a radical new direction as the quantum mechanics can reveal the significance of the picture that Sylvester discovered as a purely formal, albeit very impressive mathematical analogy." See Weyl, H.: *Philosophie der Mathematik und Naturwissenschaft*. R. Oldenburg Verlag, München, 1990, Anhang D: *Die chemische Valenz und die Hierarchie der Strukturen*, p. 351. The papers by Weyl and Rumer mentioned above were actually the primary movers toward this "radical new direction."

this apparently so wonderful agreement between spin invariants and the valence diagrams of chemistry for diatomic molecules fare with the previously demonstrated combinatoric difficulty that polyatomic molecules posed, where the number of valence lines in the chemical diagram do not equal the number of all possible spin configurations (spin invariants) along with their corresponding "pure valence states"?

Weyl showed that this apparent contradiction arises because the vectorial addition of the spins does not count all the φ invariants formed by the three or more vectors x, y, ... It only counts the linearly independent ones. This means that, as a rule, identities (linear dependencies) can occur between the φ. These are the subject of the "second fundamental law of invariant theory," according to which all linear dependencies between the monomials can be derived, in the algebraically precise meaning, from a fundamental relation (identity).

Weyl showed furthermore that in the quantum-mechanical perturbation equations, whose eigenvalues are the polyatomic molecule's energy levels, only pure states of valence that are linearly independent owing to their spin invariance couple together. For there are as many equations, hence as many eigenvalues, as there are independent invariants. Consequently, the "pure valence states," which are assignable to all possible valence-line combinations, do not in most cases correspond to stationary states of the molecule.[21] Such stationary states are rather linear combinations from "pure valence states." In general, the number of quantum states arising from perturbation is lower than the number of all possible distributions of valence lines in a molecule. Weyl points out further in his paper:

> But the quantum states in which the energy has a sharply defined value do not coincide with these valence states; they lie "in-between" them. For the eigenfunctions [...] are formed from the monomials by linear combination. In general, the number n_u of the quantum states will even be lower than the number of valence states or the possible distributions of the valence lines, because linear dependencies can exist between the monomials. These dependencies are covered by the *second fundamental law of invariant theory* according to which all such dependencies come from the one identity existing between four vectors x, y, z, t:
>
> $$[t\,x]\,[y\,z] + [t\,y]\,[z\,x] + [t\,z]\,[x\,y] = 0$$
>
> The fact of a linear dependence of valence states cannot be represented in the combinatorics of valence lines.[22]

So Weyl explicitly states that there is actually no direct correlation between the linear dependency of "pure valence states" and the valence-line diagrams used in chemistry. There is no real link between the chemist's molecule and his visual representation by means of a structural formula. Weyl does not even concede

[21] This interpretation is entirely valid for diatomic molecules and for the special class of polyatomic molecules with additive binding energies. But difficulties immediately arise with other polyatomic molecules with nonadditive energies. Cf. Heitler, W.: *Quantentheorie und homöopolare chemische Bindung*, in: *Handbuch der Radiologie* Band VI, *Quantenmechanik der Materie und Strahlung*, Teil II *Moleküle*. Akademische Verlagsgesellschaft M. B. H., Leipzig, 1934, pp. 485–586.

[22] Weyl, H.: footnote 19, p. 36.

the possibility of visualizing the mathematical property mentioned above by the chemist's valence-line diagram, i.e., portraying it by means of chemical valence structures.

Determining linearly independent invariants for polyatomic molecules was no easy matter. In 1932 the physicist Georg Rumer at the physics institute of the University of Moscow and the Karpov Institute of Physical Chemistry[23] managed to provide a general recipe by which one can draw from among the total number of possible valence states and their corresponding spin invariants an independent basis upon which to write the eigenvalue equations.[24]

If, for instance, one imagines the 4 atoms x, y, z, t in a circular arrangement, one can also depict the identity mentioned above

$$[\mathrm{t\,x}]\left[\mathrm{y\,z}\right]+\left[\mathrm{t\,y}\right][\mathrm{z\,x}]+[\mathrm{t\,z}]\left[\mathrm{x\,y}\right]=0$$

by means of the following vector diagram:

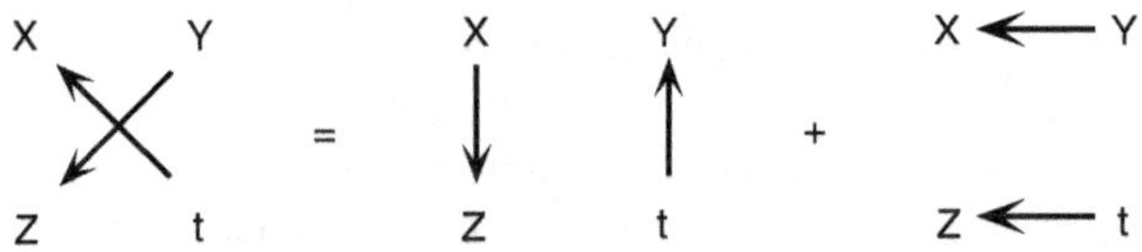

In order to take into account the signs in the equations [x y] = –[y x] and [z x] = –[x z], the valence line is rendered as a "directed valence line" (arrow). If one generally imagines the atoms distributed along a circle, all the "pure valence states" in which two valence lines cross each other can be expressed by other "pure valence states" in which no crossings occur. The "crossed" spin invariants are then derivable from a linear combination of "uncrossed" spin invariants. This means

[23]The Karpov Institute of Physical Chemistry was founded on October 4, 1918 in Moscow as a central chemistry laboratory. It was named after the first director of the Soviet chemical industry, Lev Jakovlevich Karpov. This institute enjoyed much political support because the research conducted at the institute was considered of military and economic importance for the development of the nation. During the 1930s there were eight different departments of physical chemistry. Rumer worked in the theoretical department. Since 1934 it was directed by the German pioneer of quantum chemistry Hans G. A. Hellmann (1903–1938) who had emigrated from Germany for political reasons to Moscow. His wife was Jewish. In March 1938 he was arrested in a Stalinist purge as a purported Germany spy, condemned to death in May and shot on May 29, 1938. On October 15, 1957, he and his family were exonerated. Hellmann's scientific career and personal fate is covered in Schwarz, W. H. E., Karachalios, A. u. a.: *Hans G. A. Hellmann (1903–1938) I. Ein Pionier der Quantenchemie*, in: Bunsen-Magazin 1 (1999), 10–21; *II. Ein deutscher Pionier der Quantenchemie in Moskau*, in: Bunsen-Magazin 1 (1999), 60–70.

[24]Rumer, G.: *Zur Theorie der Spinvalenz*, in: Nachrichten von der königlichen Gesellschaft der Wissenschaften zu Göttingen. Mathematisch-physikalische Klasse aus dem Jahre, 1932, pp. 337–341. Rumer's paper was presented before the society at the meeting on July 22, 1932 by H. Weyl.

that when two valence lines in a valence state cross, they can be represented as a superposition of two valence states to avoid such crossings. The result Rumer found was the following general and simple rule for finding a linearly independent basis for spin invariants: "Arrange the atoms in an arbitrary sequence along a closed convex curve (e.g., a circle) and draw in all the *valence pictures* in which no *valence lines* cross."[25] The spin invariants obtained in this way are independent of one another and exhaustive. Their total number provides a measure of the secular problem awaiting solution.

The remarkable thing about Rumer's rule is that he does not refer to "valence structures" or "structural formulas" but to "valence pictures" [*Valenzbilder*] obtained on a circle from an arbitrary yet specifically sequenced grouping of atoms – ignoring their real arrangement in the molecule – and connecting them pairwise by means of lines. Rumer noted that the "valence functions" (spin invariants) thus obtained "do not by any means depend on the arrangement of the atoms in space."[26] It follows from this that the chemist's "valence structures" or "structural formulas" in general cannot be interchanged or identified with such fictitious "valence pictures." They correspond to specific graphical formulas, namely all possible valence-line combinations, from which certain mathematical expressions (spin invariants) are obtained, in order to approximate the solution to the eigenvalue problem for calculating the molecular binding energies – or put another way, in order to come a step closer to the real solution.[27]

But because every permitted "valence picture" is correlated with a "pure valence state," it follows that a "determined energy state," representing a spin invariant, is a more general concept than "chemical molecule." It refers to an "arbitrary sequence" or localization of the atoms. The nuclear localizations of a particular molecule, however, correspond to stable equilibrium distances. So "valence pictures" and their associated "pure valence states" or spin invariants have nothing to do with possible atomic equilibrium positions, i.e., with molecular structure. Molecular structures result only secondarily in seeking nuclear configurations of minimum energy as a function of all the nuclear distances. But a "determined energy state" as a function of nuclear location may have many minimums. So it could have many "valence pictures." This situation was not yet fully understood at the time.[28] Experiment had

[25]Ibid., p. 338 (emphasis mine). By this method one obtains a complete system of independent valence states. Rumer later provided the mathematical basis in a paper written together with Teller and Weyl. See Rumer, G., Teller, E., Weyl, H.: *Eine für die Valenztheorie geeignet Basis der binären Vektorinvarianten*, in: Nachrichten von der königlichen Gesellschaft der Wissenschaften zu Göttingen. Mathematisch-physikalische Klasse aus dem Jahre, 1932, pp. 449–504.

[26]Rumer, G.: *Zur Theorie der Spinvalenz*, p. 338.

[27]The conceptual differences and specifics mentioned above play a decisive role in Hückel's criticism of Pauling's concept of resonance. See Section 3.3 below.

[28]For more details on this problem see Born, M.: *Chemische Bindung und Quantenmechanik,* in: Ergebnisse der Exakten Naturwissenschaften, 10 (1931), 387–444. Born's survey article sought to introduce physicists and chemists to Weyl's basic ideas sketched above. This purpose is spelled out in the introduction: "Weyl apparently views these results as quite insignificant side-products of his major papers on group theory, and he also published them in so brief and inconspicuous a form

meanwhile succeeded in determining the distances between the atomic nucleii of a molecule. But these internuclear distances offered a calculational simplification of the eigenvalue equations owing to the symmetry properties. Rumer pointed out: "If the spatial arrangement of the atoms has certain symmetry properties, the eigenvalue problem can be reduced."[29]

Weyl's theory, as briefly portrayed here, and Rumer's rule reduced the eigenvalue equations to a relatively simple calculation compared to previous methods. During the 1930s it was useful for calculating the molecular binding energies. These equations have a coefficient called the exchange integral that depends mainly on the distance separating two interacting atoms. Discussions of the solutions to these equations as a function of the locations of the atomic nucleii remained in most cases very complicated. So one was still a long way off from recognizing specific correlations.[30]

Despite this state of affairs during the 1930s, whether there was any direct correlation between a few of the "valence pictures" and "pure valence states" of polyatomic molecules and its structural formula remained an open question: in other words, whether a "pure valence state" existed that agreed with the molecule's formula of chemical structure and was approximately identical with its particular energetic quantum state.

Heitler was the first to focus his attention on this problem. On May 21, 1931 he submitted a paper containing his first main ideas on the topic to *Physical Review*. During that summer term Heitler was guest professor at the Ohio State University of Columbus.[31] The title of the paper was: Quantum Theory and Electron Pair Bond.[32] He showed that "for some simple examples the *stable molecular quantum state* of a polyatomic system, is *nearly identical with a state*, by which there are built so many electron pairs between each pair of atoms, as *the corresponding chemical formula*

---

that it is hard to understand. For us physicists and chemists they are, on the other hand, so important that it would seem appropriate to describe the conclusions at least in an easily comprehensible form. The proofs are not suitable for this because Slater and the author have not managed here to completely avoid the "group plague." But I suppose the practitioner will be able to suffer this if the concepts and calculational methods are offered in an understandable and usable form: the chemist conducts his experiments according to prescribed rules and recipes, so he will be willing to calculate according to prescribed rules as well."

In an effort to promote the new quantum theory of chemical bonds among chemists, Born also delivered two talks: the first on June 23, 1932, at the district affiliate in Hannover of the Verein Deutscher Chemiker and the second on February 11, 1933 in Berlin before the Deutschen Chemischen Gesellschaft. Cf. Born, M.: *Zur Theorie der homöopolaren Valenz bei mehratomigen Molekülen*, in: Angewandte Chemie 45 (1932), 6–8; *Welche Vorstellung von der chemischen Bindung vermittelt die Quantenmechanik?*, in: Angewandte Chemie 46 (1933), 179–180.

[29]Ibid.

[30]The formalisms were later simplified and presented in a more understandable form in order also to serve chemists. It was later shown, however, that the simplified version of the theory was inconsistent and a few parts were conspicuously wrong, causing this avenue to become irrelevant. Cf. Kutzelnigg, W.: *Einführung in die Theoretische Chemie*, Band 2, 1994, pp. 3–4.

[31]AHQP, Interview with Heitler, 19 March 1963, p. 2.

[32]Heitler, W.: *Quantum Theory and Electron Pair Bond*, in: Physical Review 38 (1931), 243–247.

shows."[33] Heitler at least achieved an approximative link between the theory and chemical valence theory. So it is worth citing his physical characterization of Weyl's "pure valence states" from his original text:

> There exist f_s independent wave-functions $\Psi_1, \ldots \Psi_{fs}$ which differ by the distribution p_{ab}, $p_{ac}, \ldots$ of the valence dashes between the atoms, but all have the same *sum* p of valence dashes. Weyl calls such a function a "*pure valence state,*" because the valence dashes are here strictly localized between the atoms.
>
> The wave-functions $\Phi_1, \ldots \Phi_{fs}$ belonging to the f_s different energy states $\epsilon_1, \ldots \epsilon_{fs}$ are, however, *not* identical with these pure valence states. In the zeroth order of approximation they are linear combinations of the Ψ's.
>
> $$\Phi_\iota = \sum_k c_{ik} \Psi_k$$
>
> The coefficients c_{ik} depend on the ratio of the exchange integrals, and therefore on the distances of the atoms.
>
> A pure energy state is a superposition of some pure valence states, or, in the language of the quantum mechanics the energy is not permutable by the valence – dash – configuration.
>
> It seems, that the valence-dash picture should no longer have a place in the quantum theory. We will now show, that it can again assume a certain meaning.[34]

The c_{ik} coefficients can be determined from the secular equations. In physics they signify the relative probability amplitudes for the existence of pure valence states Ψ_k, if the system is in energy state ε_i with the eigenfunction Φ_i. These probabilities are:

$$W_{ik} = |b_i \; \Phi_i \alpha_k \; \Psi_k| \; ^2,$$

where α_k and b_i are the norming factors of Ψ_k and Φ_i. If the Φ_i eigenfunctions were orthogonal and normed, it would simply be $W_{ik} = b_i^2 \; \alpha_k{}^2 |c_{ik}|^2$. This is not the case, however.

Heitler showed that in the case of some simple polyatomic molecules with linear chemical configurations approaching equilibrium, for each energy state Φ_i, just one "pure valence state" Ψ_i exists, for which the probability W_{ii} is nearly equal to one. For all the other "pure valence states" the probability is very small. They are, accordingly, repulsion states and hence correlate with structural formulas unfamiliar to chemists. For Φ_1, corresponding to energy ε_1 of the molecule's stable ground state ("stable molecular state"), the associated Ψ_1 is the "pure valence state" that almost coincides with the chemical structural formula.[35] Heitler thus demonstrated that only for certain examples is it possible to find rough agreement between quantum states of a specific energy and some "pure valence states" and that the molecule's corresponding energetic ground state can be depicted by means of a structural formula.

---

[33] Ibid., p. 243 (original English, emphasis mine).

[34] Ibid., p. 244 (emphasis in the original English).

[35] Ibid., p. 245f.

After returning to Germany Heitler wrote a survey article in which the considerations and analyses just discussed also found their place.[36] The section detailing his "near correspondence" between quantum states and the "pure valence states" closed with the hope that this correspondence perhaps be of service "as a guide for the general theory."[37]

Pauling set out in this direction but took a radical step beyond what Heitler had envisioned. He systematically identified Rumer's "valence pictures" and the associated "pure valence states" of a molecule with the structural formulas chemists had proposed, but without presenting any proof. Pauling first described Rumer's "valence pictures" and their associated "pure valence states" as "structures" and the linearly independent "pure valence states" as a "canonical set of independent structures" or "canonical vector-bond structures." What basically was involved was a formal analogy. Pauling elaborated this in his paper: The Calculation of Matrix Elements for Lewis Electronic Structures of Molecules:

> In the paper mentioned above, Rumer made the very interesting and important observation that if the orbits are formally arranged in a ring or other concave curve, and straight lines are drawn between orbits bonded together, *the structures represented by diagrams in which no lines intersect are independent*. This observation forms the starting point for the following treatment. (...)
>
> Of the various complete sets of independent structures which may be formulated, certain ones may be called "canonical" because of the relative simplicity of the calculation of matrix elements based of them. For a given order of the orbits in a ring, *the canonical set of independent structures* comprises those with no intersecting bonds; moreover, numbering the orbits in order about the ring, *the canonical vector – bond structures* for singlet states with spin degeneracy only have arrows drawn from odd to even orbits.[38]

In a second step Pauling identified "canonical structures" with the chemist's structural formulas. In the case of benzene Pauling even identified the structural formulas advanced by Kekulé and Dewar as among the above-mentioned "canonical set of independent structures," which in principle represented the linearly independent assigned "pure valence states" of the benzene molecule. He reasoned as follows:

> The benzene molecule can now be treated very simply by the Slater method, with the help of the rules formulated by one of us [39] for finding the matrix elements occurring in the secular equation. The bonds between the six eigenfunctions can be drawn so as to give the independent canonical structures shown in Fig.

[36]Cf. Heitler, W.: *Quantentheorie und homöopolare chemische Bindung*, in: *Handbuch der Radiologie* Band VI, *Quantenmechanik der Materie und Strahlung*, Teil II *Moleküle*. Akademische Verlagsgesellschaft M. B. H., Leipzig, 1934, pp. 485–586.

[37]Ibid., p. 572.

[38]Pauling, L.: *The' Calculation of Matrix Elements for Lewis Electronic Structures of Molecules*, in: JCP 1 (1933), 280–283, p. 281.

[39]Ibid.

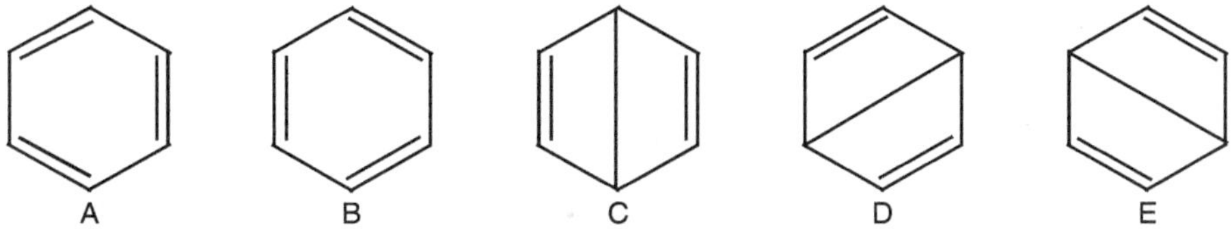

The five canonical structures contributing to the normal state of the benzene molecule

> Any other structures that can be drawn can be represented by eigenfunctions which are linear combinations of the five above. It will be seen that structures A and B are simply the two Kekulé structures, and C, D and E are three different forms of the Dewar structure. The Claus centric structure does not belong to the canonical set, but it can be represented as a linear combination of canonical structures; and the same is true of the Ladenburg structure if it be considered spread out in a plane.[40]

This identification between "canonical structures" corresponding only to particular eigenfunctions and Kekulé's and Dewar's structures opened the way for Pauling to postulate a "resonance" between various chemical "structures" and construct his own chemical theory of resonance around it.[41] Pauling's knowledge of quantum mechanics was unquestionably profound. But, as Kutzelnigg emphasized, he often looked for "quick successes, not afraid of making bold simplifications, and in doubtful cases relying more on his chemical intuition than on strict adherence to theory."[42] This situation formed the backdrop to Hückel's fundamental criticism. It was mainly launched against Pauling's conceptions of resonance and its importance in chemistry.

3.2 The First Critical Exchanges

The approximative method Pauling and Wheland used in their quantum-mechanical treatment of benzene, naphthaline and organic free radicals differed from Hückel's "first method" only in mathematical form, not in content. Pauling and Wheland

---

[40]Pauling, L.: The *Nature of the Chemical Bond. V.*, p. 364.

[41]In their paper Pauling and Wheland point out in this regard: "Hence the extra energy of the molecule resulting from resonance among the five independent structures is 1.1055α. It is interesting to see how much of this extra energy is due to resonance between the two Kekulé structures and how much is contributed by the excited structures C, D and E. A simple calculation shows that 0.9α or approximately 80 percent of the resonance energy comes from the Kekulé structures alone and only about 20 percent from the three excited structures." Ibid., pp. 364–365.

[42]Kutzelnigg, W.: *Friedrich Hund und die Chemie*, in: Angewandte Chemie 108 (1996), 629–643, p. 631. About Pauling's bold simplifications, contradictory and sometimes confused descriptions as well as the ambiguity about the physical reality of "canonical structures," see Park, B. S.: *Chemical Translators: Pauling, Wheland and their Strategies for Teaching the Theory of Resonance*, in: British Journal for the History of Science 32 (1999), pp. 21–46; Simoes, A.: *Converging Trajectories, Diverging Traditions*, p. 169.

pointed out that both methods lead to the same results.[43] All in all, there was "general qualitative agreement" between them. Quantitatively, however, as Hückel underscored in his letter to Pauling, there were a few "systematic discrepancies."[44] In Hückel's opinion, "from the methodological point of view, the work by Pauling and his school signified a substantial step forward – although, from the substantial point of view, initially a step backwards as well."[45] These papers by Pauling remained within first approximation, that is, the method that later came to be known as the Heitler-London-Slater-Pauling (HLSP) method. As has been shown in Section 2.2.2, Hückel's papers discussed how his "second method" was better suited for a quantum-mechanical treatment of the chemical behavior of benzene and other aromatic compounds. When Pauling and Wheland failed to even mention Hückel's "second method," he had to point out this omission in his letter to Pauling of December 28, 1933:

> As concerns your paper V with Wheland, I was quite astonished that you only use the Heitler-London procedure – what I call method I – and the other method – which I call method II – is not even mentioned. Do you have anything against this method?
>
> Despite your great simplification of method I, for complicated compounds method II is not only much simpler to apply than method I (which, on its own, would be no reason to give it preference, of course), but I also believe that it is better suited to the real circumstances. This mainly for two reasons: First, it yields the preferential position of 6 electrons on the ring, which method I does not reveal. Second, method I gives results for substitutions concerning the perturbed charge distribution in the ground state of benzene that cannot be brought into agreement with experiments on substitution reactions, while the contrary is the case for method II.[46]

Hückel's letter challenged Pauling to state his position not only on his second approximate method ("method II"), but also on his quantum theoretical interpretation of aromaticity and the chemical behavior of substituted benzenes.

Pauling's polite reply a few months later made it clear that he was unwilling to make any concessions concerning any superiority of the "second method." He was intransigent in his defense of the "first method."

> It was an oversight on our part not to mention your method II. (...) However, contrary to your belief, I feel strongly that method I is superior to method II.
>
> Method II for the hydrogen molecule makes the ionic part equal to the covalent part of the eigenfunction. Weinbaum has shown that is not so; the covalent part is much more important than the ionic part. I think that this result holds for more complicated molecules also, and that it shows method I to be more nearly correct than method II. But much more serious than

[43]Pauling and Wheland stated: "The results of the calculation for benzene are (...) identical with those obtained by Hückel." Cf. Pauling, L., Wheland, G.: *The Nature of the Chemical Bond. V.*, p. 365.

[44]SBPK, Papers of Erich Hückel, Box 6, Folder 5. 213, Letter Hückel to Pauling, Stuttgart (Bad Cannstatt), 28.12.1933.

[45]Hückel, E.: *Gründzüge der Theorie ungesättigter und aromatischer Verbindungen*, in: Z. Elektrochem. 43 (1937), 752–788, p. 759.

[46]SBPK, Papers of Erich Hückel, Box 6, Folder 5. 213, Letter Hückel to Pauling, Stuttgart (Bad Cannstatt), 28.12.1933.

> this is the fact that in method I complete account is taken of the Pauli exclusion principle, and in method II as applied in practice, no account whatever of the exclusion principle is taken. As a result all of the electrons in the molecule are considered to be able to pile up on the same atom. I feel that this makes the results of method II very unreliable.
> As regards the mystical number six for aromatic compounds, I think that method II and its relation to experiment is accidental. For I think the true explanation of the existence of six and five membered rings only is the influence of steric effects, principally coplanarity and the 120° degree bond angle.[47]

Pauling was quite right in pointing out in his reply that "method II" did not take the electronic interactions sufficiently into account and that it provided too large "ionic" charge densities between two bound atoms. On the other hand, "method I" neglected the "ionic" quantities altogether, which in some cases, e.g., cyclobutadiene, had fatal consequences. But it is evident from Pauling's response that he had difficulty recognizing that "method II," notwithstanding its "great simplifications," was in any way more suitable for benzene. Pauling was mistaken that "method II," with its orthogonal orbitals, does not fully obey the Pauli exclusion principle. On the contrary, as already mentioned, it does so in an extremely elegant and simple way. It is difficult to understand how Pauling could arrive at this false theoretical verdict. Pauling's rebuttal also reveals that at this early stage of his quest for an explanation of the aromatic properties of organic molecules, structural characteristics, such as benzene's valence angle of 120°, played an overly important role, quite in contrast to Hückel, who was also interested in the completeness of electron shells. Pauling and Wheland refer to this structural emphasis in their fifth paper. The section on benzene concludes as follows: "In a sense it may be said that all structures based on a plane hexagonal arrangement of the atoms – Kekulé, Dewar, Claus, etc. – play a part, with the Kekulé structures most important. It is the resonance among these structures which imparts to the molecule its peculiar aromatic properties."[48]

Pauling and Wheland were yet unable to offer a full quantum-theoretical explanation for benzene's "peculiar aromatic properties" by means of their preferred "method I." For instance, they could not account for its unusual reactive behavior upon polysubstitution. It is also conspicuous that – unlike Hückel – Pauling had set no aromaticity criteria that could serve as a practical basis for differentiation between aromatic and nonaromatic compounds and for their classification.

In autumn 1934 Hückel presented his critical views on "method I" at the International Conference on Physics in London that had been organized by the International Union of Pure and Applied Physics and the Physical Society. The main theme of the conference was: The Solid State of Matter. Sir William Bragg, member of the organizing committee, invited Hückel by letter to participate: "I am writing this letter (...) to suggest to you that your paper might deal with the work you have been doing on the benzene ring and you might add to it anything

---

[47] SBPK, Papers of Erich Hückel, Box 6, Folder 5. 115, Letter Pauling to Hückel, California Institute of Technology, Pasadena, April 17, 1934. (original English).

[48] Pauling, L., Wheland, G.: *The Nature of the Chemical Bond. V.*, p. 365.

with you think would go well with the primary purpose of your paper."[49] The topic of one of the main sessions was: The Structure of Molecules and the Ideal Lattice. Hückel delivered a talk during this session on "Aromatic and Unsaturated Molecules: Contributions to the Problem of their Constitution and Properties".[50] Hund subsequently reported on: Description of the Binding Forces in Molecules and Crystal Lattices in Quantum Theory.[51]

In London Hückel provided a summary of his papers published between 1930 and 1933. His discussion of the limitations of the classical valence-line diagram of organic chemistry for the cases of ethylene, benzene and more complex aromatic systems was unequivocal. As regards "method I," Hückel emphasized:

> Pauling and his co-workers have applied method I to a whole series of aromatic and unsaturated compounds. (...) They have further discussed the connexion between the results and the energy content of the substances. In contrast to this I considered as early as my first work on benzene, that I had been able to show that method I was unsuitable for handling the coupling of the π-electrons, since it appeared that this method leads to results which are not in agreement with experiment.
>
> Method I confines itself to the first approximation of the Heitler – London theory. Thus in the first place the objections which may be brought against the application of the method from a purely theoretical standpoint, are the same as may be raised generally against the applicability of this approximation. The method approximates to the eigenfunctions in the molecule solely through a given eigenfunction of the separate atoms, while in reality, the eigenfunctions of the higher states of the separate atoms are more or less involved in the molecular functions. This may also be expressed as a neglect of the capacity for polarization which must be taken into account in the next approximation of the Heitler – London theory. Also, no attention is given to the possibility that in a single atom there may be two electrons in the same state ("polar states" of Slater). It is very difficult to estimate in a given case to what extent the quantitative results will be affected by the neglect of these considerations. On the other hand the principal advantage of method I is that its formal scheme goes far to provide a picture of the arrangement of the valencies and shows how the classical scheme of valencies, which fails here, may be extended so that a state may be considered as arising from the superposition of different valency patterns. However, this advantage is naturally not sufficient if the quantitative results cannot be brought into agreement with experience, whereas as we shall see this will be the case for the results of method II. The failure of method I is shown by the following considerations:
>
> 1) Method I leads to results for the energy content of aromatic and unsaturated compounds which are not in agreement with experiment.
> 2) It fails to indicate the distinctive nature of rings with six π-electrons.

[49]SBPK, Papers of Erich Hückel, Box 6, Folder 5. 12, Letter Sir William Bragg to Hückel, 18 May, 1934.

[50]Hückel, E.: *Aromatic and Unsaturated Molecules: Contributions to the Problem of their Constitution and Properties*, in: International Conference on Physics, Paper and Discussions, The Physical Society, London, 1935. The University Press, Cambridge, 1935, vol. II, pp. 9–35.

[51]Hund, F.: *Description of the Binding Forces in Molecules and Crystal Lattices on Quantum Theory*, in: International Conference on Physics, Paper and Discussions, The Physical Society, London, 1935. The University Press, Cambridge, 1935, vol. II, 36–45.

> For substituted benzenes, there follows from it a disturbance of the charge distribution in the molecule, resulting from the substitutions which cannot be brought into corrrespondence with the reactive properties of the substituted benzene.[52]

Hückel then underpinned his points of criticism of the application of "method I" by Pauling and his team mentioned above with a cogent comparison between the theoretical findings and a broad range of experimental data.[53]

Hund also criticized the limits of "method I" that Hückel had emphasized in his talk with respect to the pictorial representation of the arrangement of valences of aromatic and unsaturated compounds. He discussed the interatomic binding relations in crystal lattices, metals and solids in general. In order to provide a physical explanation for their properties, Hund argued, the quantum theory could be introduced at various points of the theoretical considerations. He distinguished three explanatory stages:

> As a *first and rather crude step* we consider atoms as given and assume forces between them, repulsive on account of the impenetrability, attractive on account of the electrical structure of the atom. (...) This step in the theory did not explain the special properties of different substances, nor, above all, the rules of chemistry.
>
> At a *more advanced stage* the individual properties of atoms, such as the existence and energy of positive or negative atomic ions, are deduced from the quantum theory of the atom. *The aggregation of matter is then explained as being due to the electrostatic forces between the ions*, so that *the explanation is classical and pictorial.* To this stage belong Kossel's explanation of heteropolar chemistry and Born's theory of the atomic lattice. The chief points left unexplained are homopolar molecules, metals, and certain solid insulators like diamond.
>
> In *the third stage*, the quantum theory of the molecule and the crystal lattice, electrons are regarded as moving in a system of a few nuclei (as in molecules) or of many nuclei (as in crystal lattices) *in a non-classical and non-pictorial manner* by the methods of the *(non-relativistic) quantum mechanics* initiated by Heisenberg and Schrödinger (1925–26). This stage of development leaves most of the results of the earlier stages untouched; however, it adds some new results, and *explains the properties of aggregated substances which are not explicable on classical, pictorial lines, viz. homopolar chemistry* and solid atomic lattices.[54]

Hückel's and Hund's emphasis on a non-pictorial, "non-classical" representation of the binding relations of a few chemical compounds is, in my opinion, of special significance in the history of quantum chemistry of the 1930s. The overwhelming majority of chemists, including the pioneering figure Pauling, unquestioningly constructed their conceptions from the two-dimensional structural formulas, or pictorial

---

[52]Hückel, E.: *Aromatic and Unsaturated Molecules*, p. 15. (original English).

[53]Ibid., pp. 15–20. Hückel had conveyed some of his points of criticism to Pauling prior to the conference in London in a letter dated July 25, 1934, requesting "a prompt reply." No documentation has yet been found on whether Pauling wrote in reply.

[54]Hund, F.: *Description of the Binding Forces in Molecules and Crystal Lattices on Quantum Theory*, in: International Conference on Physics, Paper and Discussions, The Physical Society, London, 1935. The University Press, Cambridge, 1935, vol. II, 37–38 (emphasis mine.)

models of molecules in three-dimensional space.[55] As Simoes and Gavroglu have aptly described, Pauling used a "pragmatic approach," a semi-empirical method in fundamental agreement with the classical chemical conceptions, to interpret by quantum theory a few problematic aspects in chemistry.[56] Indeed, Mary Jo Nye has recently demonstrated that Pauling generally preferred an "architectural approach to chemistry," that emerged from the "pencil and paper tradition of structural chemistry."[57] Hückel and Hund offered an opposing alternative, a novel conceptual approach that contrasted starkly to the traditional visual thinking of the organic chemist. An important element of this conceptual approach, as Hückel stressed, was that it was called for not only on a theoretical basis but on an experimental one as well.

Slater also took part in the discussions following the talks by Hückel and Hund at the London conference. He explicitly gave his stamp of approval to their evaluation of method II:

> Professors Hückel and Hund have pointed out the two methods which have been used for computing interatomic forces, the first depending on electron pairs, the second on molecular orbitals, and have indicated that the second gives results in better agreement with experiment. I wish to state that I am in entire agreement with this conclusion, and that I believe that the second method, in the approximation to which it can be conveniently carried, is better than the first in a corresponding approximation.[58]

His laconic reference to Pauling's application of "method I," without going into further detail, was:

"The calculations of Pauling, for instance, seem to make quite unwarranted use of the theory."[59] Hückel continued to uphold this insinuated criticism by Slater of Pauling's illegitimate use of the theory. In the years that followed, his attacks on Pauling's preferred method I were strident, particularly against his interpretations of the underlying mathematics and his concept of resonance.

---

[55]From among the abundant literature on this aspect, we refer to the following important paper and further references there. Francoeur, E.: *Molecular Models and the Articulation of Structural Constraints in Chemistry*, in: *Tools and Modes of Representation in the Laboratory Sciences*, ed. by Ursula Klein. Kluwer Academic Publishers, Dordrecht/Boston/London, 2001, pp. 95–115; Hoffmann, R. und Laszlo, P.: *Darstellungen in der Chemie – die Sprache der Chemiker*, in: Angewandte Chemie 103 (1991), 1–16.

[56]Cf. Gavroglu, K., Simoes, A.: *The Americans, the Germans and the Beginning of Quantum Chemistry* (1994).

[57]Nye, M. J.: *Paper Tools and Molecular Architecture in the Chemistry of Linus Pauling*, in: *Tools and Modes of Representation in the Laboratory Sciences*, edited by Ursula Klein. Kluwer Academic Publishers, Dordrecht/Boston/London, 2001, pp. 117–132.

[58]Slater, J. C.: *Discussion on the Structure of Molecules and of the Ideal Lattice*, in: International Conference on Physics, Paper and Discussions, The Physical Society, London, 1935. The University Press, Cambridge, 1935, vol. II, p. 53.

[59]Ibid.

3.3 "Special Portion of Energy" Versus "Resonance Energy"

A year after the conference in London, Hückel received a letter from Germany's "wellspring of organic chemistry": Friedrich Richter (1896–1961). This editor of Beilstein's handbook on organic chemistry[60] had taken an interest in the new ideas in quantum theory about chemical bonding and wanted to ask Hückel a question regarding his papers on the benzene issue. He wondered what specific experimental data Hückel had employed in his calculation of the energy content of aromatic compounds and what his views were on the subject. Richter was convinced that "at the moment the only part that an organic chemist can reasonably play in quantum mechanical research is evaluating the experimental foundations."[61] His inquiry into these "experimental foundations" also touched on Pauling's research. In Richter's opinion, Pauling "evidently makes liberal use of the thermochemical data, without being aware, or at least sufficiently aware, of how extremely uncertain the foundations of thermochemistry still are today."[62] Richter's critique shook a supporting pillar of Pauling's conceptual framework on the nature of chemical bonding, namely, his additivity postulate of the energies of covalent bonds that proved so fruitful for his chemical theory of resonance.[63]

Hückel soon responded to Richter's questions[64] and critique of Pauling's school in a letter dated October 31, 1935:

---

[60]SBPK, Papers of Erich Hückel, Box 6, Folder 5. 117, Letter Richter to Hückel, Berlin, 28 Oktober, 1935.

[61]Ibid.

[62]Ibid.

[63]Pauling formulated this postulate in his paper: The Additivity of the Energies of Normal Covalent Bonds, which he submitted on 9 May 1932 to the *Proceedings of the National Academy of Sciences of the United States of America*: "In the wave function representing the bond between unlike atoms A and B, the ionic term A^+B^- and A^-B^+ will occur with the same coefficient, of the order of magnitude of those for A:A and B:B, if the two atoms have the same degree of electronegativity. We propose to call such a function a normal covalent bond wave function, and the bond a normal covalent bond; and to make the postulate that *the energies of normal covalent bonds are additive*, that is, A:B = 1/2 (A:A + B:B), where the symbols A:B, etc., mean the energies of the normal covalent bonds. This postulate requires that the energy change for a reaction such as 1/2 A_2 + 1/2 B_2 = AB involving only normal covalent substances with single bonds be zero." Cf. Pauling, L., Yost D. M.: *The Additivity of the Energies of Normal Covalent Bonds*, in: Proceedings of the National Academy of Sciences of the United States of America 18 (1932), 414–416 (original emphasis). Details about the importance of this postulate in the development of Pauling's research agenda on the nature of chemical bonding and his theory of resonance are given in Weininger, S. J.: *Affinity, Additivity and the Reification of the Bond*, in: *Tools and Modes of Representation in the Laboratory Sciences*, edited by Ursula Klein. Kluwer Academic Publishers, Dordrecht/Boston/London, 2001, pp. 237–251; Simoes, A.: *Converging Trajectories, Diverging Traditions* (1993), pp. 156–158.

[64]Hückel enclosed with his letter to Ritter a separatum of his London talk in which he laid out in detail the experimental data used to calculate the energy content of the various aromatic compounds.

> In my view this school goes far too far in applying its calculational scheme. For the benefit of *formal elegance* it forgets the limitations imposed on the validity of the approximation employed, owing to the mathematical constraints on the permitted solutions inherent in the approximative method. In my view they apply it to illegitimate cases. [...] I have the impression that, on the basis of a kind of dogmatic prejudice for their method, the Pauling school have the tendency to squeeze areas of experimental observation into a scheme that from an unbiased point of view one would have to expect to be too narrow for such types of data. I find it very interesting to hear that you as an organic chemist are inclined to take a similar attitude toward this research. Quite in contrast to Pauling, I do believe I have taken great pains to point out the limitations of my theory and have constantly scrutinized them against the experimental data from every conceivable angle.[65]

Other complaints Hückel implied in his letter to Richter about the conceptual and calculational schemes employed by the "Pauling school" became more detailed over the next 2 years. 1937 Hückel published an overview article in the *Zeitschrift für Elektrochemie* entitled "Outlines of the theory of unsaturated and aromatic compounds."[66] It surveyed the state of the art of the theory in a language that chemists were able to understand. The main target of Hückel's attacks was Pauling's concept of resonance.

His point of departure was the characteristic of unsaturated and aromatic compounds resulting from his own theory that "– unlike the case of two atoms, each with an electron of the second kind[67] – *the number of possible valence structures no longer agrees with the number of different states of determined energy*; and that even *a state of determined energy can no longer be assigned a specific valence structure*."[68] Hückel named benzene as a classical example. He had already shown in his first paper on the benzene problem by means of "method II"[69] that a state "of determined energy (and particularly its ground state) cannot be assigned a specific structure renderable by means of dashes."[70]

Pauling, on the other hand, had interpreted his calculational method in his fifth paper in such a way that the composition of benzene be approximated from "canonical structures" corresponding to the structural formulas proposed by Kekulé and Dewar. Using the phenomenon of resonance in mechanics as an analogy, Pauling referred to a "resonance" between these structures and an associated "resonance energy."[71]

---

[65] SBPK, Papers of Erich Hückel, Box 6, Folder 5. 214, Letter Hückel to Richter, Stuttgart, 31 Oktober, 1935.

[66] Hückel, E.: *Grundzüge der Theorie ungesättigter und aromatischer Verbindungen*, in: Z. Elektrochem. 43 (1937), 752–788; 43 (1937), 827–849.

[67] With it are meant the $[p]_h$-Elektronen or π-Elektronen.

[68] Hückel, E.: *Grundzüge der Theorie ungesättigter und aromatischer Verbindungen*, p. 758 (original emphasis).

[69] For more details see Section 2.2.2 above.

[70] Hückel, E.: *Grundzüge der Theorie ungesättigter und aromatischer Verbindungen*, p. 758.

[71] As Pauling's concept of resonance has already been thoroughly discussed by Gavroglu, Park and Simones, this section will only be concerned with Hückel's critique against Pauling. Cf.

With an eye to Pauling's vocabulary, Hückel commented: "The expression 'resonance' insinuates [an] incomplete analogy pointing to an interaction between separate valence structures through the coupling of vibratable mechanical systems of the same frequency."[72] In Hückel's opinion, the Kekulé structures could only be applied "in formal analogy to two swinging pendula that are uncoupled and have the same frequency."[73] In reality, Hückel argued, "both structures exist simultaneously and so neither of them can be ascribed a specific energy, nor therefore a frequency, just as in the case of coupled pendula."[74] He then continued with an explanation of the limitations of the analogy between mechanical and quantum-mechanical resonance.

> Unlike the case with pendula, it is not possible here to have an oscillatory state that would be a superposition of symmetrical and antisymmetrical oscillations and in which "suspension" could occur in such a way that at one time only one Kekulé structure (corresponding to the swing of a pendulum) could occur and then, after a half period of suspension, only the other Kekulé structure (corresponding to the swing of the other pendulum) and in the interim both structures. Because in quantum theory every state [i.e., "stationary states"] is assigned a determined energy and hence a frequency; unlike with pendula, it cannot be composed of two frequencies at the same time. Kekulé's oscillation hypothesis is therefore inappropriate as well.[75]

Contrary to Pauling's first conception, Hückel could not regard the Kekulé structures or the oscillation hypothesis as corresponding to physical reality. Hückel was of the radical opinion that, even disregarding the incomplete and misleading analogy between mechanical and quantum-mechanical resonance, "the expression 'resonance' between the structures and the expression 'resonance energy' were indeed misleading and therefore to be discarded."[76] Hückel proposed that the expression "mesomerism" should be used instead of resonance, "in conformance with

---

Park, B. S.: *Chemical Translators: Pauling, Wheland and their Strategies for Teaching the Theory of Resonance*, in: British Journal for the History of Science 32 (1999), 21–46; *Computations and Interpretations: The Growth of Quantum Chemistry, 1927–1967*, Dissertation, John Hopkins University, Baltimore, Maryland, 1999; Simoes, A.: *Coverging Trajectories, Diverging Traditions* (1993); Simoes, A., Gavroglu, K.: *Issus in the History of Theoretical and Quantum Chemistry, 1927–1960*, in: *Chemical Sciences in the 20th Century*, edited by Carsten Reinhardt. Wiley-VCH, Weinheim, 2001; Mosini, V.: *A Brief History of the Theory of Resonance and its Interpretation*, in: Studies in History and Philosophy of Modern Physics 31B(4) (2000), 569–581.

[72]Hückel, E.: *Grundzüge der Theorie ungesättigter und aromatischer Verbindungen*, p. 763.

[73]Ibid.

[74]Ibid.

[75]Ibid.

[76]Ibid., p. 767.

Christopher Kelk Ingold,[77] and recently Fritz Arndt[78] and Bernd Eistert,[79] who also

---

[77]Ingold, C. K.: *Principles of an Electronic Theory of Organic Reactions*, in: Chemical Reviews 15 (1934), 225–274; *Mesomerism and Tautomerism*, in: Nature 133 (1934), 946–947. On Ingold's career in science and his contributions to the concept of "mesomerism" as well as to structures and mechanisms in organic chemistry as a whole, see Shoppee, C. W.: *Christopher Kelk Ingold*, in: Biographical Memoirs of Fellows of the Royal Society 18 (1972), 349–411; Nye, M. J.: *From Chemical Philosophy to Theoretical Chemistry*, Chap. 8, *Reaction Mechanisms: Christopher Ingold and the Integration of Physical and Organic Chemistry, 1920–1950*, pp. 196–225; *Remodeling a Classic: The Electron in Organic Chemistry, 1900–1940*, in: *Histories of the Electron, The Birth of Microphysics*, edited by Jed Z. Buchwald and Andrew Warwick. The MIT Press, Cambridge, Massachusetts, 2001; Schofield, K.: *The Development of Ingold's System of Organic Chemistry*, in: Ambix 41 (1994), 87–107; Brock, W. H.: *The Norton History of Chemistry*, Chap. 14; Saltzmann, M.: *C. K. Ingold's Development of the Concept of Mesomerism*, in: Bulletin for the History of Chemistry 19 (1996), 25–32.

[78]Arndt, F., Eistert, B.: *Über den "Resonanz" – und "Zwischenstufen" – Begriff bei organischen Substanzen mit mehrfachen Bindungen und die Elektronenformeln*, in: ZPC-B 31 (1935), 125–131.

Fritz Arndt, born in Hamburg on July 6, 1885, was one of the few organic chemists in Germany occupied with theoretical issues on chemical bonding. From 1919 he was supernumerary professor of organic chemistry at Breslau, receiving tenure in 1928, only to be dismissed in 1933 as a "non-Aryan." Soon afterwards, the Turkish state officially invited German professors seeking asylum to come to Istanbul. Arnst was one of 19 German professors whose scientific exchanges with Turkey date back to after World War I, when the German and Ottoman Empires were allies. He received an appointment to the newly founded "Istanbul Universitesi" on August 1, 1933, where he built up a laboratory of organic chemistry. He and his students examined tautomerizable systems, proving the effectiveness of methylation with diazomethane. These analyses led to a distinction between static and dynamic acidity.

Eistert was a long-time collaborator of Arndt, maintaining contacts with him after his emigration. They copublished a number of papers until 1941. One was the paper just mentioned, in which the "resonance concept" was described in terms of a theory of "intermediate stages" that Arndt had derived 11 years earlier out of purely chemical considerations. In his paper "On dipyrylenes and on the binding relations in pyrone ring systems," published in the *Berichte* of 1924, he found that the chemical behavior of pyrones could not be correctly expressed either by formula I or by the "bonding isomers" zwitterion formula II. It was an "intermediate state" between the two extremes represented by the formulas:

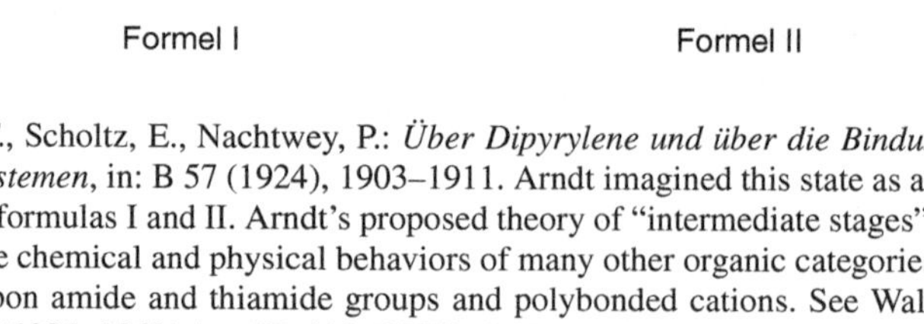

Formel I Formel II

Cf. Arndt, F., Scholtz, E., Nachtwey, P.: *Über Dipyrylene und über die Bindungsverhältnisse in Pyron-ringsystemen*, in: B 57 (1924), 1903–1911. Arndt imagined this state as an "intermediate stage" between formulas I and II. Arndt's proposed theory of "intermediate stages" proved fruitful in explaining the chemical and physical behaviors of many other organic categories of substances, such as the carbon amide and thiamide groups and polybonded cations. See Walter, W., Eistert, B.: *Fritz Arndt (1885–1969)*, in: CB 108 (1975), I–XLIV; Campaigne, E.: *The Contributions of*

used this designation."[80] Hückel preferred this term because it seemed to express particularly well "the states of the molecule *lying*, as it were, *between the fictitious states* that correspond to the canonical structures."[81] Pauling, for his part, opted against using "mesomerism": "Because the resonating system does not have a structure intermediate between those involved in the resonance, but a structure which is further changed by resonance stabilisation I prefer not to use the word 'mesomerism', suggested by Ingold, for the resonance phenomenon."[82]

Hückel suggested furthermore that instead of referring to "resonance" energy (W_r) one use the neutral expression: "special portion" of energy, and denote this extra portion as W_s rather than W_r.[83] This way confusion about its meaning could better be avoided. In a later article with the title "On the modern theory of

---

Fritz Arndt to Resonance Theory, in: Journal of Chemical Education 36 (1959), 336–339; Üstün, A.: *Zweites Vaterland – deutsche Chemiker im türkischen Exil*, in: Nachrichten aus der Chemie 51 (2003), 152–155; Burk, L. A.: Fritz *Arndt and his Chemistry Books in the Turkish Language*, in: Bulletin for the History of Chemistry 28 (2003), 42–52.

[79]Bernd Eistert was born on November 9, 1902 in the regional Slask town of Ohlau near Breslau. 1922 he took up the study of chemistry at the University of Breslau. The director and chair of the institute there was the inorganic chemist Heinrich Biltz. Julius Meyer was extraordinary professor of inorganic chemistry and Fritz Arndt was a private lecturer in the faculty. Eistert took his doctorate under Arndt 1927 on chain extension in carboxylic acids. On Arndt's recommendation he became assistant of the complex chemist Paul Pfeifer in Bonn on November 1, 1928, and in 1929 he was employed at the main laboratory of BASF in Ludwigshafen. He remained in touch with Arndt and continued to work on such theoretical problems as tautomerism and mesomerism, the chemism of Claisen condensation, etc. Eistert's interest in theory prompted exchanges with Erich Hückel about Hückel's survey article "Grundzüge der Theorie ungesättigter und aromatischer Verbindungen" mentioned above as well as on Eistert's own successful book "Tautomerie und Mesomerie". Cf. Regitz, M., Heydt, H., Schank, K. und Franke, W.: *Bernd Eistert (1902–1978)*, in: Chemische Berichte 113 (1980), XXIX–LVIII.

[80]Hückel, E.: *Grundzüge der Theorie ungesättigter und aromatischer Verbindungen*, p. 764. Hückel cited in his article the following papers by Ingold, Arndt and Eistert: Ingold, C. K.: *Principles of an Electronic Theory of Organic Reactions*, in: Chemical Reviews 15 (1934), 225–274, insbesondere S. 250–252; Arndt, F., Eistert, B.: *Über den Chemismus der Claisen-Kondensation*, in: B 69 (1936), 2381–2398.

[81]Ibid. (emphasis mine).

[82]Pauling, L.: *The Nature of the Chemical Bond and the Structure of Molecules and Crystals: An Introduction to Modern Structural Chemistry*. Cornell University Press, New York, 1939, p. 10, footnote 1. Hückel then criticized the previously quoted passage from Pauling's book, emphasizing the contradictory meaning of the term "structure": "Here in one and the same sentence Pauling uses two meanings for the word 'structure': once with the meaning a diagram that is assigned to a specific function that can (but does not have to!) be used as one of the initial functions for the perturbation calculation; the other time, to designate the real *state* of the molecule; for such functions are not changed anymore. Only specific linear combinations are sought, in particular those that lead to the lowest energy in the perturbation calculation." Cf. Hückel, E.: *Zur modernen Theorie ungesättigter und aromatischer Verbindungen*, in: Z. Elektrochem. 61 (1957), 866–890, p. 873. (original emphasis).

[83]Ibid.

unsaturated and aromatic compounds,"[84] Hückel repeated what he considered as crucial arguments against the use of the term "resonance." In his opinion, he wrote, what was basically involved was not a process in which a "resonance" arose between different "structures" as in classical physics, but merely "an analogy with a mathematical calculation procedure – a purely mathematical formalism," for solving the secular problem. This must not be confused with a real physical phenomenon.[85] He continued:

> I went into such length about this not just to quibble about words but because I had to see, time and again, that many chemists attach inappropriate meanings to the designations "resonance," "resonance energy," and "resonance stabilization." For example, that at one instant the molecule is in a "state" depicted by the one "structural formula," at another instant, in a "state" depicted by another. One is misled to this interpretation by the expression resonance together with the *formal mathematical analogy of coupled pendula*. Or the idea that "resonance" or so-called "resonance stabilization" were a *physical process*. Or else the idea that a compound could react "in the sense of one structural formula or another"; this is inaccurate simply because such "structural formulas" do not describe any molecular state. It would therefore be better to speak of "*line diagrams*" rather than of *structures*.[86]

Hückel then drew Weyl's physical interpretation of "pure valence states" and Rumer's assigned "valence pictures" into his discussion of this subject:

> On the basis of the general theory on permutation groups and on algebraic invariants, Heitler and Rumer as well as Weyl could prove the following.
> For n atoms (n = even), each with one π electron (we shall limit ourselves to this case here), there are
>
> $$\frac{n\,!}{\left(\frac{n}{2}+1\right)!\;\left(\frac{n}{2}\right)!}$$
>
> such linearly independent functions belonging to $S = 0$ [i.e., the quantum number of the total spin], thus, e.g., two for $n = 4$ (butadiene), five for $n = 6$ (benzene), 42 for $n = 10$ (naphthaline), etc.
> Each of these functions may be assigned a "*line diagram*" in the following way: One arranges the atoms – independently of their location in space! – in an arbitrary but determined sequence on a circle and link them pairwise by a line; then each of these "*line diagrams*" corresponds to one of the functions belonging to S = 0. These functions are not linearly independent of each other, however. There are rather as many linearly independent functions as there are possible line diagrams without intersecting lines. The functions assigned to these are practically (but not necessarily!) applied to the perturbation calculation.[87]

Hückel stressed that such "line diagrams," which Pauling called "structures" or "canonical structures," had absolutely "nothing to do with molecular structure"[88]

---

[84] Hückel, E.: *Zur modernen Theorie ungesättigter und aromatischer Verbindungen*, in: Z. Elektrochem. 61 (1957), 866–890.

[85] Ibid., pp. 872–873.

[86] Ibid., p. 873 (emphasis mine).

[87] Ibid., p. 871 (emphasis mine).

[88] Ibid., p. 871, footnote 19.

and consequently could not represent the ground state of a molecule. In Hückel's view, which he shared with Weyl and Rumer, they only represented the way in which the "antisymmetrical product functions"[89] can be composed in order to obtain the functions for the total spin ($S = 0$), for use in the perturbation calculation.[90]

Heitler's retrospective remarks about the expression "resonance," which agreed with Hückel's objections to Pauling's resonance concept of the 1930 s, is an appropriate close to this chapter. In his article on the path-breaking era of quantum chemistry from 1967, Heitler wrote: "The expression 'resonance' (...) is hardly adequate. Resonance means an exchange of energy. The various valence structures are degenerate unperturbed states, due to various possible spin orientations. Between the two Kekulé structures there is no more a resonance than between the three states of a p level."[91] Heitler's remark after the fact shows how difficult it was for the scientists of the day to wean themselves from classical visualizations and tackle true considerations of quantum theory.

[89]According to Slater, they can be written in the form of determinants. Thus they are also known as "Slater functions" or "Slater determinants."

[90]Ibid.

[91]Cf. Heitler, W.: *Quantum Chemistry: The Early Period*, in: International Journal of Quantum Chemistry 1 (1967), 13–36, p. 34, footnote 4.

Chapter 4
Linus Pauling's Breakthrough to the Theory of Aromatic Compounds and Hückel's Reaction

In the mid-1930s Wheland and Pauling set out to find a quantum theoretical basis for Ingold's general electronic theory of organic reactions. A brief outline of Ingold's main concepts will provide the necessary background for an evaluation of their contribution and Hückel's discussion of it.

4.1 Introduction

As has been shown in Section 2.2.4, in his article on benzene from 1931 Hückel attempted to explain the guiding influence one substituent has on subsequent substitutions as its effecting a change in the electronic charge distribution on the benzene ring among the 6 $[p]_h$ electrons. He also sought an interpretation according to quantum theory. Hückel made the following assumption on the basis of his quantum-mechanical calculations: a weakening of the charge density of the $[p]_h$ electrons, that is, a rise in positive charge among the carbon atoms on the benzene ring, facilitates acceptance of a second substituent; a strengthening of the charge density impedes the occurrence of a second substitution.

Meanwhile, between 1926 and 1933 some English authors developed a general electronic theory of organic reactions that had yet to be founded on quantum theory.[1] Ingold chose an opposite assumption to Hückel's as a basis for the influence of a change in electronic density (charge density) on secondary substitutions. A greater charge density among the $[p]_h$ electrons along the benzene ring facilitates secondary substitution.[2] Besides taking into account the charge disturbance among the carbocyclic atoms, which Hückel described as an "inductive effect," Ingold also considered a disturbance of opposite effect called the "mesomeric effect."[3] The

[1]Cf. Nye, M. J.: *From Chemical Philosophy to Theoretical Chemistry* 1993, Chap. 8, *Reaction Mechanisms: Christopher Ingold and the Integration of Physical and Organic Chemistry, 1920–1950*, pp. 196–226.

[2]Cf. Ingold, C. K.: *Principles of an Electronic Theory of Organic Reactions,* 1934.

[3]Cf. Ingold, C. K.: *The Principles of Aromatic Substitution from the Standpoint of the Electronic Theory of Valency*, in: Recueil des Travaux Chimiques 48 (1929), 797–812; *Significance of*

A. Karachalios, *Erich Hückel (1896–1980)*, Boston Studies in the Philosophy of Science 283, DOI 10.1007/978-90-481-3560-8_4,

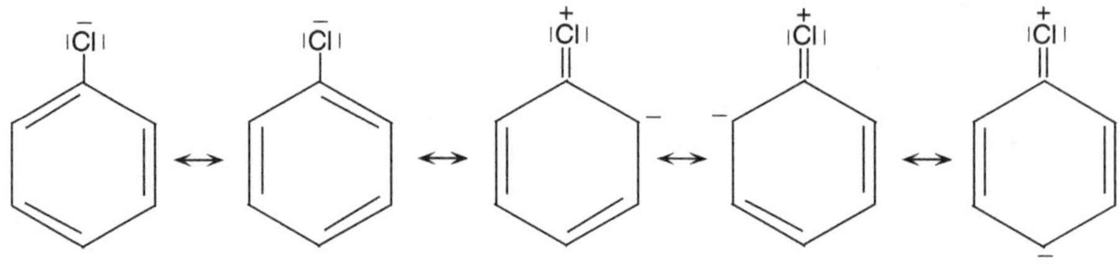

Fig. 4.1

charge of the available electrons of a first substituent, e.g., of a halogen $X{=}C1$, partly travels along the benzene ring. (See Fig. 4.1)

Consequently, the "mesomeric effect" counteracts the "inductive effect," because an accumulation of $[p]_h$ electrons in ortho-para positions corresponds to the "polar structures" (IIa) and (IIb). Leslie E. Sutton had already demonstrated by a series of measurements of the dipole moments, feasible since 1926, that the mesomeric effect was stronger than the inductive effect.[4] This experimental verification encouraged Ingold to take the "mesomeric effect" alone into account. This assumption was also supported by experimental data on the secondary substitution of halogen benzenes in ortho and para positions.

Finally, Ingold took into account what he called the "inductomeric effect." It later became known as the "polarizability effect."[5] This effect only takes place under the influence of an approaching substituent ("attacking group"). It is not the charge distribution of the free molecule that governs the reaction but the charge distribution existing during (or just before) the reaction as a result of the influence of the approaching substituent. Ingold assumed an approaching substituent was on the look-out for electrons ("electron seeking"). So the result of the effect is considered for the charge distribution of the $[p]_h$ electrons attached to the carbon atom being approached by the substituent. Ingold called the change in charge distribution by approaching substituents the "inductomeric effect."[6]

Thus, there were altogether three significant effects in Ingold's electronic theory of organic reactions for a satisfactory explanation of the mechanisms of aromatic secondary substitutions.

Tautomerism and of the Reactions of Aromatic Compounds in the Electronic Theory of Organic Reactions, in: Journal of the Chemical Society (London) (1933), 1120–1127.

[4]Sutton, L. E.: *The Significance of the Differences between the Dipole Moments of Saturated and Unsaturated Substances*, in: Proceedings of the Royal Society of London (A) 133 (1931), 668–695.

[5]Cf. Nye, M. J.: *From Chemical Philosophy to Theoretical Chemistry* (1993), pp. 208–210.

[6]It is noteworthy that Arndt and Eistert included this effect caused by external disturbance in their concept of mesomerism. Cf. Arndt, F., Eistert, B.: *Über den "Resonanz" – und "Zwischenstufen" – Begriff bei organischen Substanzen mit mehrfachen Bindungen und die Elektronenformeln*, in: ZPC-B 31 (1935), 125–131.

4.2 Avowal of the Molecular Orbital Method

In July 1935 Wheland and Pauling submitted their paper, Quantum Mechanical Discussion of Orientation of Substituents in Aromatic Molecules, to the *Journal of the American Chemical Society.*[7] It adopted Ingold's three effects in constructing a new theory based on quantum mechanics to explain substitution reactions on substituted benzenes. This chapter was also Wheland and Pauling's answer to Hückel's fitting criticism about their application of "method I." A few months earlier, during the London conference, Hückel had pointed out that the limitations of the method prevented Pauling from arriving at a correct quantum-theoretical explanation of polysubstitutions of the benzene molecule.[8]

Besides applying Ingold's three effects (which they referred to as inductive, resonance and polarizability effects), in their introduction Wheland and Pauling also drew on the experimental data from dipole moment measurements to refute Hückel's methodology for explaining substitution reactions of aromatic compounds by quantum theory.

> [Hückel] limited his calculations, however, to the inductive effect, and neglected the resonance effect as well as the polarization of the molecule by the attacking group. As a result of these over-simplifications, he was led to make incorrect physical and chemical assumptions in order to obtain a rough correlation between his calculations and experiment. (...)
>
> The error in Hückel's treatment lies not in the quantum mechanical calculations themselves, which are correct as far as they go, but in the oversimplification of the problem and in the incorrect interpretation of the results. Consequently it has seemed desirable to us to make the necessary extensions and corrections in order to see if the theory can lead to a consistent picture. In the following discussion we have found it necessary to consider all of the different factors mentioned heretofore: the resonance effect, the inductive effect, and the effect of polarization by the attacking group.[9]

This time, Wheland and Pauling surprisingly tried to calculate the simultaneous effects just mentioned by using not their preferred "method I" (valence bond method) but "method II" (molecular orbital method). They justified this unexpected about-face as follows: "There are two principal methods available for the quantum mechanical treatment of molecular structure, the valence bond method and the molecular orbital method. In this paper we shall make use of the latter, since it is more easily adapted to qualitative calculations."[10] Because of its emphasis on "polar states," Wheland and Pauling were reluctant to accept that their favored "valence bond method" was inapplicable So they applied a "molecular orbital method"

[7] Wheland, G. W., Pauling, L.: *A Quantum Mechanical Discussion of Orientation of Substituents in Aromatic Molecules*, in: JACS 57 (1935), 2086–2095.

[8] See Section 3.2.

[9] Wheland, G. W., Pauling, L.: *A Quantum Mechanical Discussion of Orientation of Substituents in Aromatic Molecules*, 1935, p. 2088.

[10] Ibid.

that was basically an extension of "method II" that Hückel had used for his own "quantum theory of induced polarities."

Wheland and Pauling calculated the molecular electronic states, hence the charge density, of the eight π electrons (six $[p]_h$ electrons and two from substituent X) in the fields of the seven atoms (six carbocyclic atoms and one substituent X). For this they characterized the atomic fields of the π electrons by certain parameters δ_κ denoting the deviation of the potential field of one electron at atom κ from the potential field of one of the carbon atoms in benzene. More precisely put, the quantities δ_κ gave the deviation of the eigenvalue of a π electron at atom κ against one at a carbon atom in undisturbed benzene, measured as a multiple of the "resonance integral" β, characteristic of this method, for two π electrons of neighboring carbon atoms on the ring.[11] It was not yet possible to say anything more specific about the change of the eigenfunction for larger δ_κ and about the resonance integral. For this reason, Wheland and Pauling neglected these deviations for larger values of δ_κ. They pointed out: "We have attempted to estimate values of the δ's for the different kinds of atoms (...), but we have not been able to arrive at any very consistent figures. Our procedure was to determine the values which would bring the calculated resonance energies into agreement with the empirical ones."[12] Assuming these simplifications and neglecting electronic interaction, Wheland and Pauling employed the molecular orbital method to calculate the simultaneous influence of Ingold's three effects on the distribution of charge among the π electrons for various aromatic compounds.[13]

On 18 December 1936 Hückel submitted a new paper criticizing Wheland and Pauling's interpretation of substitution reactions of aromatic compounds by quantum theory.[14] In his thorough discussion of the theses advanced by the coauthors regarding the three effects, Hückel arrived at the following results:

First, he emphasized that as yet no calculation of the charge disturbances of the carbon atoms on the ring of a halogen benzene had been completely successful, because both he himself and Wheland and Pauling had neglected to calculate the change in the distribution of charges among the σ electrons.

Second, Hückel pointed out that it was not certain that the mesomeric effect outweighed the inductive one, because in his opinion it was not even clear whether

[11] Ibid., p. 2089.

[12] Ibid., p. 2090.

[13] Wheland and Pauling summarized the result of their investigations as follows: "Using the method of molecular orbitals, a quantitative discussion of the charge distribution in aromatic molecules undergoing substitution reactions is carried out, taking into consideration the inductive effect, the resonance effect, and the polarizing effect of the attacking group. It is shown that, with reasonable values for the parameters involved, the calculated charge distributions for pyridine, toluene, phenyltrimethylammonium ion, nitribenzene, benzonitrile, furan, thiophene, pyrrole, aniline, phenol, naphthalene, and the halogen benzenes are in qualitative agreement with the experimental results regarding position and rate of substitution." Ibid., p. 2095.

[14] Hückel, E.: *Kritische Betrachtungen zur Theorie der Substitutionsreaktionen an substituierten Benzolen*, in: ZPC-B 35 (1937), 163–192.

the assumptions of Ingold's model were even correct. In supporting this assertion Hückel critically analysed experimental findings obtained by Sutton.

Third, Hückel argued that the polarizability effect on the substitutability, which was comparatively small against the other two effects, must be viewed as purely hypothetical.

Hückel concluded unyieldingly in summary: "But in any case it is not possible to arrive here at a truly satisfactory theory by the methods hitherto employed."[15] Moreover, he radically contended that the Wheland–Pauling attempt to interpret by quantum theory Ingold's conceptions could "not be regarded as successful in any way."[16]

In this context it is relevant to mention that until the mid-1930s the specifics on the true mechanisms governing substitution reactions remained unknown.[17] That was why Hückel considered it impossible to decide between the two hypotheses, his own or the one by Ingold that Wheland and Pauling were attempting to interpret quantum-theoretically.[18] Later developments were to prove his hypothesis wrong, however. Both experiment and theory came out in favor of Ingold's theory.

4.3 A New Concept for Describing Aromaticity

In autumn 1934 at the International Conference on Physics in London, Hückel closed his talk on the constitution and properties of aromatic and unsaturated compounds with the clairvoyant statement: "The treatment of further properties, such as magnetism and light absorption, is a task for further research."[19]

Two years later, in a paper bearing the title The Diamagnetic Anisotropy of Aromatic Molecules, Pauling set out in a direction Hückel had originally indicated – there is no documentation on whether Hückel had any direct influence on this decision.[20] It addressed the idea of a "ring current" in conjugated cyclic molecules. It prompted further investigations of aromatic compounds using magnetic techniques and the development of magnetic criteria for describing aromaticity. Pauling started from the assumption "that the p_z electrons (one per aromatic carbon atom) are free to move from carbon atom to adjacent carbon under the influence of the impressed [magnetic] fields."[21]

---

[15]Ibid., p. 188.

[16]Ibid., p. 192.

[17]Ibid., pp. 165–169.

[18]Ibid., p. 169. Cf. Hückel, E.: *Grundzüge der Theorie ungesättigter und aromatischer Verbindungen*, pp. 844–845.

[19]Hückel, E.: *Aromatic and Unsaturated Molecules: Contributions to the Problem of their Constitution and Properties*, in: International Conference on Physics, Paper and Discussions, The Physical Society, London, 1935. The University Press, Cambridge, 1935, vol. II, p. 35. For further details about Hückel's talk, see Section 3.2.

[20]Pauling, L.: *The Diamagnetic Anisotropy of Aromatic Molecules*, in: JCP 4 (1938), 673–677.

[21]Pauling, L.: *The Diamagnetic Anisotropy of Aromatic Molecules*, 1938, p. 673.

Most organic molecules have no permanent magnetic moments and are therefore weakly *diamagnetic*. This means that they have *negative molar-magnetic susceptibilities*. Magnetic susceptibility describes the behavior of a compound in an external magnetic field. When a magnetic field of the strength H_0 is applied, what is called *molar magnetization* M is induced in the molecules, where $M = \chi_M \cdot H_0$. χ_M denotes the *molar-magnetic susceptibility*. Compounds with $\chi_M < 0$ are called *diamagnetic*, and those with $\chi_M > 0$ are designated as paramagnetic. Diamagnetism is generated by the motions of the structure of π electrons. They produce small magnetic fields of opposite direction to the applied magnetic field. The magnitude of this effect in aromatic compounds depends on the size of the orbit the electron is on. Most diamagnetic molecules are *anisotropic*. This means that the diamagnetic susceptibilities along the three main perpendicular magnetic axes have different values.

The crystals of a number of aromatic compounds including benzene, naphthaline and anthracene manifest large diamagnetic anisotropy. In 1933 K. S. Krishnan and his team managed to demonstrate that in these compounds the relation between the magnetic and crystal axes varies with the orientations of the molecules in the crystal. They also found out that the magnitude of magnetic susceptibility along these magnetic axes depends on the magnetic susceptibility along the molecule's main axis.[22]

Elaborating on Paul Ehrenfest's earlier theory, C. V. Raman and Krishnan suggested a qualitative explanation of diamagnetic anisotropy.[23] Pauling developed this interpretation further, giving it a quantitative basis. By means of a first-order perturbation formula Pauling showed that the contribution of an electron in a cylindrically symmetric field around the z axis to the magnetic susceptibility is given by the equation[24]

$$\chi = -\left(\frac{Ne^2}{4mc^2}\right)\left(\rho^2\right)_{au}$$

In this equation, χ is the magnetic susceptibility and $(\rho^2)_{au}$ is the root mean square of the electron's distance away from the z axis. With this formula Pauling calculated the diamagnetic anisotropy of benzene and a series of other aromatic fused ring systems that agreed well with the measured values.[25]

After World War II, Pauling's assumption that the "p_z electrons" circulate along the ring was used to explain the screening of aromatic protons in the NMR spectrum. The reason for this screening was that the ring current induced by the cyclic motion of the electrons affected the ^{1}H NMR spectrum.[26] With the development of NMR

[22]Cf. Garratt, P., Vollhardt, P.: *Aromatizität*, 1973, pp. 28–30.

[23]Raman, C. V., Krishnan, K. S.: *Magnetic Double-Refraction in Liquids. Part I – Benzene and its Derivates*, in: Proceedings of the Royal Society of London A 113 (1927), 511–519.

[24]Pauling, L.: *The Diamagnetic Anisotropy of Aromatic Molecules*, 1936, p. 674.

[25]Ibid., p. 677.

[26]Pople, J. A.: *Proton Magnetic Resonance of Hydrocarbons*, in: JCP 24 (1956), p. 1111.

spectroscopy, the concept of aromaticity was once again applied to a property of materials, this time to a substance's magnetic anisotropy.[27]

I mention here the Irish physicist and crystallographer Kathleen Lonsdale (1903–1971), who had already analysed the anisotropy of various compounds in the solid state. In her recent paper on Magnetic Anisotropy and Electronic Structure of Aromatic Molecules she pointed out with reference to Pauling's idea of a ring current: "The idea that certain electrons can move in this way is implicit in the theory of molecular orbitals, as applied by Hückel to aromatic and unsaturated molecules."[28] In his first paper on the benzene problem, Hückel had indeed interpreted the eigenfunctions of the [p]h electrons of benzene physically as "circulating electronic waves [...], that as a consequence of the periodicity in F are modulated in this periodicity."[29] However, before the appearance of Pauling's paper on the diamagnetic anisotropy of aromatic compounds (besides the allusion during the conference in London mentioned above), Hückel never referred, either directly or indirectly, for instance to the direction (toward the right or left) of the "circulating electronic waves" or to possible relations between the circulation of the [p]h electrons and the magnetic properties of aromatic compounds observed in experiment. A year after the appearance of Pauling's paper, however, the situation changed.

In his survey article "Outlines of the theory of unsaturated and aromatic compounds," to be discussed in more detail in Chapter 5, Hückel's verdict on Pauling's paper was positive: "Pauling has succeeded in treating the special diamagnetic behavior of aromatic compounds according to quantum theory on the basis of the concept of molecular electronic states." But he was no less stringent a critic. He added: "The validity of the method Pauling has applied still awaits more detailed theoretical foundation, however."[30] He continued: "Pauling calculates, for instance, as though the electrons of the second kind [i.e., the [p]h electrons] could circulate freely in the benzene ring, whereas in reality while circulating they must pass through the potential barriers between the C atoms."[31]

Interestingly enough, the theoretical foundation Hückel was missing came from another pioneer of quantum chemistry: Fritz London, who in 1937 was working in the Institut Henri Poincaré at the University of Paris. At that time he was working on superconductivity and diamagnetism, which topics he had taken up since his emigration from Germany in 1933.[32]

---

[27]See Section 2.2.1, footnote 121 (bibliography on aromaticity).

[28]Londale, K.: *Magnetic Anisotropy and Electronic Structure of Aromatic Molecules*, in: Proceedings of the Royal Society of London 159 (1937), 149–161.

[29]Hückel, E.: *Quantentheoretische Beiträge I*, p. 241.

[30]Hückel, E.: *Grundzüge der Theorie ungesättigter und aromatischer Verbindungen*, 1937, p. 837, footnote 2.

[31]Ibid.

[32]For more details see Gavroglu, K.: *Fritz London a Scientific Biography*, Chap. 3, *Oxford and Superconductivity bzw.* Chap. 4, *Paris und superfluidity*.

London provided a basis in quantum theory for Pauling's postulated free motion of the "p_z electrons" and at the same time extended the Hückel–Mulliken "molecular orbital theory" to the presence of a magnetic field.[33] He then presented Pauling's calculations in a simple and straightforward manner, "stating that the *aromatic compounds shall behave like supraconductors*."[34]

It seems that Hückel took the opportunity of Pauling's article to elaborate on his original physical interpretation of the $[p]_h$ electrons as "circulating electronic waves." At any rate, he wrote in his paper from 1937: "The electrons can have a *circulating direction* around the ring."[35] Going a step further he asserted:

> Accordingly, the molecular eigenfunctions here are not standing waves but circulating (modulated) waves. For a given energy determined by a wave circulating toward the right, there is a wave of the same energy circulating toward the left. The two associated states are what we call, mutually "degenerate." However, it is characteristic of the new quantum theory, and just for this new theory, that other states of determined energy exist to which only one eigenfunction is attributable and which do not have any direction of circulation.[36] These states are, as we say: "nondegenerate" or "simple".[37]

With this additional physical interpretation of the $[p]_h$ electrons, Hückel was able to offer a clearer description of the electronic states of benzene. He imagined the six molecular electronic states of benzene's $[p]_h$ electrons as follows:

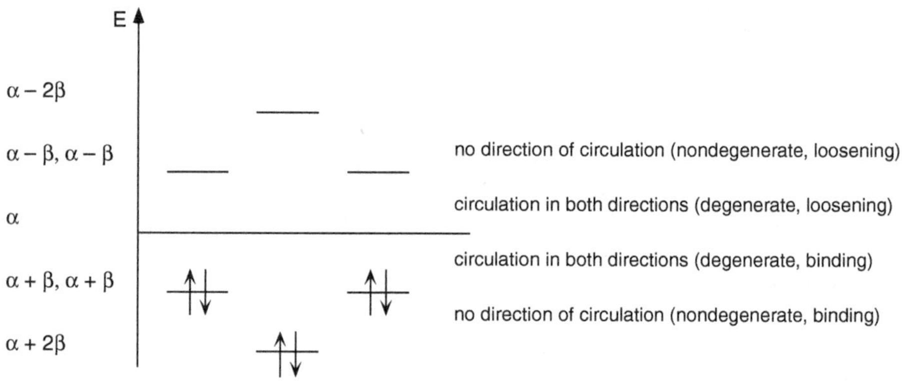

Hückel's new type of depiction shows clearly that no direction of circulation is attributable to two of the six electrons, while the four others are paired, each with

[33] Cf. London, F.: *Théorie quantique du diamagnétisme des combinaisons aromatiques*, in: Comptes Rendus 205 (1937), 28–30.

[34] London, F.: *Supraconductivity in Aromatic Molecules*, in: JCP 5 (1937), 837–838. (emphasis in the original English). See also London, F.: *Théorie quantique des courants interatomiques dans les combinaisons aromatiques*, in: Journal de Physique et le Radium 8 (1937), 397–409.

[35] Hückel, E.: *Grundzüge der Theorie ungesättigter und aromatischer Verbindungen*, 1937, p. 777.

[36] Hückel added the comment that eigenfunctions belonging to such states are real.

[37] Hückel, E.: *Grundzüge der Theorie ungesättigter und aromatischer Verbindungen*, 1937, p. 777.

their counterpart circulating in opposing senses at equal angular momentum. Thus, Hückel explained, in total there is no angular momentum and therefore no magnetic moment as a consequence of the electronic circulation. Because, furthermore, "the spins are oppositely paired, benzene is therefore diamagnetic."[38] Hückel thus integrated Pauling's idea of the cyclic motion of the p_z electrons into his earlier conceptions.

[38]Ibid., p. 778.

Chapter 5
Hückel's Efforts to Disseminate His Theory and Its Reception

Shortly after giving his talk in autumn 1934 at the International Conference on Physics in London, Hückel sought to introduce the results of his research on unsaturated and aromatic compounds elsewhere in Europe. On April 30, 1935 Hückel delivered a talk in French before the *Société de Physique* in Paris. The title he chose was: "Aromatic and unsaturated compounds. Theoretical inquiry into their constitution and properties."[1] Describing his own efforts as clearly as he could, he also discussed his points of criticism against Pauling's theoretical and mathematical approach. How his Parisian audience responded seems not to have come down to posterity. Hückel had been invited to give a series of other talks abroad, as he later recalled: "A tour through Holland was planned, there was an invitation to Bologna, but then came Hitlerism and that was the end of that."[2]

Besides going abroad, Hückel was mainly preoccupied with transmitting his ideas at home. His native country had been under National Socialist rule since 1933. Hückel also advertised the importance of the new theoretical advances that quantum physics could offer chemistry. For instance, at the "national meeting of German chemists" that took place in Munich between July 7 and 11, 1936, Hückel gave a talk that went to the crux of the matter: "The importance of the new quantum theory for chemistry."[3]

In this talk Hückel emphasized above all two important aspects of the new quantum theory for chemistry. First, it provided new experimental methods of research. Hückel was referring to spectral analysis based on the new wave and quantum mechanics. Raman spectroscopy is an example as well as observations of electron interference aiding determinations of the atomic arrangements inside molecules. Spectroscopy gave information about possible rotational, oscillatory and electronic states in molecules and their energies. It also gave information about products of dissociation and dissociation energies and other molecular characteristics. Second,

[1] Hückel, E.: *Composés aromatiques et non saturés. Recherches théoriques concernant leur constitution et leurs propriétés*, in: Le Journal de Physique et le Radium 6 (1935), 347–358.

[2] *Chemiker im Gespräch: Erich Hückel*, p. 185.

[3] Hückel, E.: *Die Bedeutung der neuen Quantentheorie für die Chemie*, in: Z. Elektrochem. 42 (1936), 657–662.

A. Karachalios, *Erich Hückel (1896–1980)*, Boston Studies in the Philosophy of Science 283, DOI 10.1007/978-90-481-3560-8_5,

quantum theory could base on physical principles conceptions about molecular composition and reactivity that chemists had already developed. It also was capable of deepening and extending existing knowledge, providing answers to problems that had proved too difficult for former chemical approaches. Hückel presented unsaturated and aromatic compounds as an example. He also pointed out the necessity of a collaboration between chemistry and physics, as physics was bound to serve "practical chemistry" well in future. Likewise, concerning some "training in physical mathematics" for chemists, which in his opinion was otherwise not easily acquired.[4] In his Munich talk he explained the difficulties chemists encountered in attempting to apply the theory effectively as due to the complexity of the systems of interest to chemists as well as due to the "abstractness" and "unintuitiveness" of the new theory:

> Penetrating into the new theory is not easy, for two reasons. First, the theory differs from earlier theories in physics in that the atomic processes can no longer be described by motions of particles or waves in normal space. For this a space at a higher dimension is needed, what is called configurational space; and concepts must be used that have no direct intuitive meaning in space and time. For this reason the theory has the reputation of being unintuitive and abstract. But what we deem intuitive is a matter of familiarity; and abstraction is required for every intellectual effort. [. . .]
>
> But I cannot refrain from mentioning that the lack of interest in the theory among chemists, it seems to me, does not absolve them completely from responsibility about how we have been applying and representing the theory; the more so, since the chemical knowledge of some theoreticians who have been applying the theory to chemical issues occasionally has been less than sufficient.[5]

Hückel thus pointed his finger at German chemists for their apathy about the theory just as much as at theoretical physicists for their unfamiliarity with chemical problems and conceptions. Basically, he was missing true interdisciplinary communication between chemists and theoretical physicists in the Germany of his day. A few months before his talk in Munich, Hückel had discussed this subject with the editor of Beilstein's handbook on organic chemistry, Friedrich Richter. In his previously cited letter to this editor[6] Hückel outlined his point of view:

> The overall relations between theoretical physics and organic chemistry is indeed quite unsatisfactory right now. You are entirely right to say that a theoretical treatment of problems of interest to organic chemists demands such a thorough familiarity with the mathematical methods of quantum physics that it is not possible for the organic chemist to work theoretically on these problems or to form a judgment about the merit of theoretical research in this area. On the other hand, in my view, the majority of theoreticians working on these problems must be blamed for not trying hard enough (or perhaps, like the organic chemist but vice versa, is not in a position) to "think chemically," if I may put it that way.

---

[4]Ibid., p. 662.

[5]Ibid., pp. 661–662.

[6]See also Section 3.3.

> However, in my view, the subtlety and individuality of chemical phenomena pose the greatest barriers toward application of theoretical physics to organic chemistry in a manner that would make it fruitful for organic chemists working on such problems.[7]

At the end of his talk, Hückel implicitly suggested bridging this gap in interdisciplinary thinking among theoretical physicists and chemists with professorships in this new field, taking Anglosaxon countries as role models: "In recognition of the importance of the new theory in chemistry, much effort is being made in England and America to apply this theory to chemical issues. Indeed, there already are professorships devoted to the subject."[8] But thirty years had to go by before this vision won widespread support among chemists in Germany.[9]

Shortly after the conference in Munich, during the winter of 1936, Hückel spoke before the German chemical society in Berlin at the instigation of Richter, who was a member of its speaker selection committee.[10] It was presumably at Richter's suggestion that he chose the topic: "The application of quantum mechanics to the benzene problems."[11] Hückel reminisced that he accepted a number of other speaking engagements as well. By his own assessment, the one he delivered in Bonn was a "miserable talk," blaming "the irritating automatic blackboard mechanism."[12] Besides, he suspected that the majority of the participants, who were chemists, hardly even understood his talks. Hückel wrote in his autobiography about these

---

[7]SBPK, Papers of Erich Hückel, Box 6, Folder 5. 214, Letter Hückel to Richter, Stuttgart, 31 Oktober, 1935.

[8]Hückel, E.: *Die Bedeutung der neuen Quantentheorie für die Chemie*, p. 662.

[9]It was only during the 1960s that German academia established its first professorships in theoretical chemistry. The new centers were the University of Göttingen under the directorship of W. Bingel (from 1964), Free University in Berlin with Ernst Ruch (from 1966), Stuttgart with Heinzwerner Preuß (from 1967), and Munich with Ludwig Hofacker (from 1968). At the same time, the subject theoretical chemistry was introduced into the curriculum. The council at the University of Frankfurt founded in 1966 a "center for theoretical chemistry" with two chairs and three assistantships. Hermann Hartmann had been the original initiator. In 1960 he had formed a small group of theoreticians with the financial support of the Deutsche Forschungsgemeinschaft (DFG). Through its focus program on theoretical chemistry, this national research fund provided substantial support for research in quantum chemistry, considerably furthering dissemination of the discipline. Hartmann, H.: *Theoretische Chemie – Förderung einer modernen Wissenschaft in Frankfurt*, Sonderdruck, Vittorio Klostermann, Frankfurt am Main 1968; Karachalios, A.: *Die Entstehung und Entwicklung der Quantenchemie in Deutschland*, in: Mitteilungen der Gesellschaft Deutscher Chemiker Fachgruppe Geschichte der Chemie 13 (1997), 163–179; Peyerimhoff, S.: *Theoretische Chemie: Quantenchemie*, in: *Forschung in der Bundesrepublik Deutschland*, hrsg. von Christoph Schneider, Verlag Chemie, Weinheim, 1983, pp. 611–619; *The Development of Computational Chemistry in Germany*, in: Reviews in Computational Chemistry, vol. 18, edited by Kenny B. Lipkowitz and Donald B. Boyd, John Wiley and Sons, Inc., Wiley – VCH, 2002, pp. 257–291.

[10]SBPK, Papers of Erich Hückel, Box 6, Folder 5. 117, Letter Richter to Hückel, Berlin, 20.12.1935 und 30.10.1936.

[11]SBPK, Papers of Erich Hückel, Box 6, Folder 5. 117. Richter argues in favor of this suggested topic.

[12]*Chemiker im Gespräch: Erich Hückel*, p. 185.

events: "I tried my best with these talks, it was terribly difficult, and perhaps it was no easier trying to understand them."[13] Notwithstanding Hückel's difficulties in transmitting his ideas in a comprehensible form to chemists, only a few physical chemists, and perhaps the exceptional chemist, were willing to delve into his highly abstract papers replete with integrals, equations and matrices. For this reason and intent on attracting the interest of chemists in Germany to recent quantum-theoretical research on unsaturated and aromatic compounds, Hückel finally decided to write a clear overview of this area in a language that the chemical public could understand.

He decided to visit his brother Walter in the fall of 1936 for a number of weeks. A year before, the National Socialists at the Culture Ministry had transferred Walter from Greifswald to Breslau.[14] His purpose was to discuss with Walter the chemical issues of pertinence to his planned survey.[15] This was not the first time that the Hückel brothers had been engaged in intense discussions about these issues. As Erich Hückel wrote in the acknowledgment to his survey article, that the editors of the *Zeitschrift für Elektrochemie und angewandte physikalische Chemie* received on June 20, 1937,[16] he had been benefiting from "a fruitful exchange of ideas" with his brother Walter since starting to work on unsaturated and aromatic compounds.[17] He later emphasized that "all these [...] papers [would] probably never have come about without my communications with my brother. My brother was a walking Beilstein [handbook] or a walking *Zentralblatt* for me."[18] Richter also played a crucial role in the drafting of the survey article. According to Hückel's acknowledgments, his "encouraging invitations provided the stimulus for my writing" the paper.[19]

Hückel set himself the goal of presenting a clear overview of the state of the art of the theory of unsaturated and aromatic compounds. His primary goal was to make the basic ideas and the relevant concepts of quantum theory comprehensible. Hückel also described the extent to which quantum mechanics reached beyond classical chemical conceptions. This necessarily led Hückel to outright confrontation with Pauling's approximative method and physical interpretation. From Breslau Hückel wrote to Lars Onsager, his friend dating back to his Zurich period, who had in the meantime emigrated to America:

---

[13]Ibid.

[14]Cf. Neidlein, R.: *Walter Hückel*, in: CB 113 (1980), I–V, p. III. Concerning Walter Hückel's transfer, Erich Hückel mentions in his autobiography: "But [Walter] was "transferred" from his professorship in Greifswald against his will, because a P[arty] C[omrade] wanted to have this position and then got it as well." Hückel, E.: *Ein Gelehrtenleben* (1975), p. 140.

[15]Hückel W.: *Memoiren*, pp. 439–440.

[16]The paper concerned is the previously cited survey "Grundzüge der Theorie ungesättigter und aromatischer Verbindungen".

[17]Ibid., p. 849.

[18]The abstracts published in the *Zentralblatt für Mathematik* were a common source of reference. *Chemiker im Gespräch: Erich Hückel*, p. 185.

[19]Hückel, E.: *Grundzüge der Theorie ungesättigter und aromatischer Verbindungen*, 1937, p. 849.

> I will now be here in Breslau for a while visiting my brother Walter, who was transferred here about 1 1/2 years ago [...]. I want to do the preliminary work here for another paper on aromatic compounds, which will probably have a quite negative streak. For I absolutely cannot befriend myself with Pauling's papers, because according to my view much in them is done too brashly, and too uncertain and far-reaching conclusions are drawn. I intend to get to the bottom of the matter for once.[20]

The result of this focus on Pauling's papers was, as we have seen in Section 3.3, a quite radical assessment concerning in particular the physical interpretation of quantum-mechanical resonance between the Kekulé structures of benzene. In this survey article Hückel explicated that there was no similarity between quantum-mechanical resonance and the resonance phenomena of classical mechanics.[21] Thus he consciously rejected interpreting resonance between valence structures in quantum-mechanical considerations of the benzene molecule in terms of the visual concepts of classical mechanics. The reason Hückel gave was that the new quantum mechanics deprived of meaning such concepts as "electron orbits" and mechanical frequencies of electrons in a chemical bond, replacing them with the new, unintuitive concept of "electronic states." The result of Hückel's quantum-mechanical analysis of the C=C double bond in unsaturated and aromatic compounds was that one had to distinguish between "two kinds of electronic states" of determined energy: σ electrons and π electrons,[22] which are characterized by specific eigenfunctions. That was why in quantum mechanics one ought rather to refer to a molecule's energetic state, which is not visualizable in every case by the valence structures of classical organic chemistry.[23]

Besides going into the problem of depicting chemical matters by quantum mechanics in a visualizable or intuitive way, Hückel also pointed explicitly to the then prevailing limitations of the theory. It fell short as much in explaining the correlations between constitution and reactivity as it did regarding the correlations between constitution and color. In both cases Hückel thought an extension of the methodology was necessary. Hückel regarded the ground state of a molecule as not the only factor responsible for a compound's color or reactivity: its excited states were equally important. He thus perspicuously anticipated future developments:

> Hence the absorption of light occurs by stimulating a molecule, i.e., by a transition from the ground state into excited states. The *excitation energy* is crucial for the level of the absorption frequency; the *transition probability* for a transition into the appropriate excited state is essential for the absorption coefficient of an absorption frequency. Most previous attempts to interpret the observed relations between constitution and color must be regarded

---

[20] Universitetsbiblioteket – NTNU, Lars Onsager arkiv, Box 2.55, Hückel an Onsager, Breslau, 30. IX, 1936.

[21] Hückel, E.: *Gründzüge der Theorie ungesättigter und aromatischer Verbindungen*, 1937, p. 763.

[22] Hückel called them "electrons of the first kind" and "electrons of the second kind."

[23] Hückel, E.: *Gründzüge der Theorie ungesättigter und aromatischer Verbindungen*, 1937, pp. 753–754. It is also noteworthy that Hückel already described such topics in the part he contributed to the first edition of his brother Walter's two-volume work on the theory of organic chemistry. Erich's part was titled "The development and fundamental principles of quantum theory." Cf. Hückel, W.: *Theoretische Grundlagen der Organischen Chemie*, Band II, pp. 66–75.

> as fundamentally inadequate, however, because they did not take these things into account, seeking instead a direct correlation between the constitution, i.e., the electronic arrangement in the ground state, and the color.[24]

Hückel just mentioned this idea as a thought-provoking impulse. Mulliken was among those who continued to develop the theory of light absorption of organic compounds between 1938 and 1940. He published a series of papers under the title Intensities of Electronic Transitions in Molecular Spectra.[25] Applauding Hückel's survey article as "excellent," Mulliken carried the ideas "implicitly" introduced in Hückel's work forward, particularly in his third paper.[26]

After the first part of Hückel's survey article appeared in *Zeitschrift für Elektrochemie*, the chief editor Georg Grube had the idea of publishing it in the form of a "special brochure, as it were, as a communication of the Deutschen Bunsen Gesellschaft."[27] As full professor of physical chemistry and electrochemistry at the polytechnic in Stuttgart, Grube had already pulled his weight toward arranging a lectureship there for Hückel. The managing director of the Bunsen Society, Alexander Schweitzer, then wrote to Debye, who had been awarded the Nobel prize in chemistry a year before for his contributions on molecular structure. In his letter dated September 16, 1937 he asked "whether on the basis of the part that has recently appeared you could already judge whether this would be a paper suitable for the purpose mentioned. [...] Would you, as the case may be, be willing to prepare an report on this paper?"[28] Debye had meanwhile read not only the first of Hückel's articles but its continuation as well and imparted his opinion in a letter from November 1, 1937:

> A considerable amount of work has been invested in Hückel's paper and I believe that there is no one else in Germany who could treat the question of the application of quantum theory to chemistry even nearly as well as Hückel. There is furthermore no question in my mind that chemists will have to concern themselves somewhat more than before with the conceptions of quantum theory. At the same time, it is equally clear to me that true advantage can only be gained if it is possible to eliminate as much of the entire mathematical apparatus as possible and, at the same time, to condense the ideas into as clear and intuitive a picture as possible in order to be able to foster a proper feel for the fundamentals of wave mechanics. From this point of view I applaud Hückel's article most enthusiastically. I do

---

[24] Hückel, E.: *Gründzüge der Theorie ungesättigter und aromatischer Verbindungen*, 1937, p. 848.

[25] Mulliken, R.: *Intensities of Electronic Transitions in Molecular Spectra. I. Introduction*, in: JCP 7 (1939), 14–20; *II. Chargetransfer Spectra*, in: JCP 7 (1939), 20–34; *III. Organic Molecule with Double Bonds*, in: JCP 7 (1939), 121–135; *IV. Cyclic Dienes and Hyperconjugation*, in: JCP7 (1939), 339–352; *V. Benzene*, in: JCP 7 (1939), 353–356; *VI. Molecular Refractivities of Organic Compounds*, in: JCP 7 (1939), 356–363; *VII. Conjugated Polyenes and Carotenoids*, in: JCP 7 (1939), 364–373; *VIII. Odd-Numbered Conjugated Polyene Chain Molecules and Organic Dyes (with Notes on Optical Anisotropy and Raman Intensities)*, in: JCP 7 (1939), 570–572; *IX. Calculations on the Long Wave-Length Halogen Spectra*, in: JCP 8 (1940), 234–243; *X. Calculations on Mixed-Halogen, Hydrogen Halide, Alkyl Halide, and Hydroxyl Spectra*, in: JCP 8 (1940), 382–395.

[26] Mulliken, R.: *III. Organic Molecule with Double Bonds*, pp. 121 und 123, footnote 6.

[27] BALM, A. Schweitzer an P. Debye, Stuttgart, September 16, 1937.

[28] Ibid.

> not mean to say that this article attains the ideal I have imagined. But it is a major step in this direction. [...]
>
> In my opinion, the Bunsen Society would be doing theoretical chemistry in its modern form a service if it made Hückel's article easily accessible and helped disseminate it as widely as possible.[29]

Debye's evaluation convinced the editor with the result that Hückel's article appeared in the following year as a separate brochure.[30]

On November 18, 1937 Hückel received a letter from the director of the Kaiser Wilhelm Institute in Heidelberg, Richard Kuhn (1900–1967), one of the leading German chemists of the 1930s – and supporter of Hitler's aggressive war policy:[31]

> In the name of the Kaiser Wilhelm Institute of Medical Research I would most respectfully like to invite you to deliver a talk in our colloquium on the new theoretical basis for treating aliphatic and aromatic compounds. Not just I, but a number of other gentlemen are most vibrantly interested in your recent analyses, judging particularly from the survey that you published in the Zeitschrift für Elektrochemie, and from an oral discussion we hope to gain inspiration in addition to that afforded by perusal of your paper.[32]

Hückel actually did follow Kühn's invitation and presented a talk before his colleagues in Heidelberg a few months later, on April 25, 1938.[33] The precise title of the talk is not known. What is certain, though, – if we may believe Hückel's recollection – is that Hückel once again gave a "miserable talk."[34] Hückel's deficient communication skills – quite the opposite of Pauling's charisma – undoubtedly had negative consequences on the proliferation and acceptance of his path-breaking research. On the whole, Hückel's quantum-theoretical papers made little impact on the chemical community in Germany. Besides the reasons already discussed, the dominance of preparative organic chemistry in Germany was also significant. On the basis of classical structural theory of aromatic compounds it had been gathering major discoveries, particularly in the area of dyes, since the 19th century.[35]

---

[29] BALM, Debye an Schweitzer, November 1, 1937.

[30] Hückel, E.: *Grundzüge der Theorie ungesättigter und aromatischer Verbindungen*. Verlag Chemie, Berlin, 1938. Vgl. auch Jaenicke, W.: *100 Jahre BUNSEN-Gesellschaft*, Steinkopff, Darmstadt, 1994, p. 124.

[31] For details on Kuhn's scientific achievements and his political views during the National Socialist regime and his collaboration on projects of importance to the war effort, see Deichmann, U.: *Flüchten Mitmachen Vergessen*, Kap. 6 *Naturstoffchemische und biochemische Forschung im internationalen Vergleich vor dem Hintergrund der Entwicklungen bis 1933*, Kap. 7.4 *Richards Kuhns Forschung während des Zweiten Weltkriegs – Befruchtungsstoffe, Sulfonamide und Gaskampfstoffe*.

[32] SBPK, Papers of Erich Hückel, Box 6, Folder 5. 110, Letter Kuhn to Hückel, Heidelberg, 18 November 1937.

[33] SBPK, Papers of Erich Hückel, Box 6, Folder 5. 110, Letter Kuhn to Hückel, Heidelberg, 12 February 1938.

[34] *Chemiker im Gespräch: Erich Hückel*, p. 185.

[35] Cf. Reinhard, C.: *Forschung in der chemischen Industrie: Die Entwicklung synthetischer Farbstoffe bei BASF und Hoechst, 1863 bis 1914*. Technische Universität Bergakademie Freiberg: TU Bergakademie, Freiberg 1997; Reinhardt, C., Travis, A. S.: *Heinrich Caro and the Creation of Modern Chemical Industry*. Kluwer Academic Publishers, Dordrecht, 2000.

This success resulted in a relative neglect of theoretical aspects and a retrenchment against other disciplines, such as physics, that had developed possible applications in chemistry. This, in turn, had the effect that such subfields of theoretical chemistry as quantum chemistry lay fallow in Germany. The physical chemist Walther Jaenicke remembered:

> I see the absence of one segment of theoretical chemistry, primarily the one concerned with quantum-theoretical foundations of chemical bonding, less as a consequence of National Socialism than as a consequence of a lack of receptiveness by organic chemists toward new developments. Although quantum chemistry was founded in Germany on the work of Hund and Hückel, it was not cultivated here. [...] For decades all chairs in chemistry in Germany were occupied by organic chemists. Erich Hückel, for instance, had hardly any chance of being heard.[36]

Organic chemists at universities, just as in industry, generally favored the empirical results and approaches used in their application-oriented and synthetic laboratory research. Theories with complicated mathematical apparatus and conceptualizations were relegated onto the side-lines.[37]

Under the sway of National Socialism with its autarkic and militaristic policies, the German chemical industry with its promising advances in practical applications assumed a stronger role in chemical research than before.[38] As a consequence, the areas of pure theory in research and teaching were pushed even further aside. The forced emigration of a majority of the pioneers in quantum chemistry, notably Walter Heitler, Fritz London and Hans Hellmann, as a result of the racist policies of the National Socialists substantially impeded the development of this field as well.[39] Thus Hückel lost potential counterparts in discussions about his ideas, although it remains an open question whether they would have welcomed his quantum-chemical approach, since they had mostly developed different kinds of approximative methods for interpreting chemical issues.

Soon after Hückel's papers on the benzene problem were published, they received some notice abroad. In 1935 the physical chemist and Italian pioneer

---

[36] Walther Jaenicke's interview with Ute Deichmann, Erlangen, 18.08.1996, quoted in: Deichmann, U.: *Flüchten, Mitmachen, Vergessen*, p. 158. Cf. Jaenicke, W.: *100 Jahre BUNSEN-Gesellschaft*, Kap. 1.37 *Physik, Chemie und nationalsozialistische Ideologie*, pp. 120–124.

[37] Jeffrey, A. J.: "*Academic-Industrial Relations and Chemical Education in Germany, 1919–1939*" (draft paper presented to Fourth International Conference of the CHMC, "Industrial-Academic Relationships in the Chemical and Molecular Sciences", October 4, 2002).

[38] Stokes, R. G.: *Von der I. G. Farbenindustrie AG bis zur Neugründung der BASF (1925–1952)*, in: *Die BASF. Eine Unternehmensgeschichte*, Verlag C. H. Beck, München 2002.

[39] Deichmann, U.: *Flüchten, Mitmachen, Vergessen*, Kap. 3.4.4 Theoretische Organische Chemie; Jost, W.: *The First 45 Years of Physical Chemistry in Germany*, in: Annual Review of Physical Chemistry 17 (1966), 1–14; Kutzelnigg, W.: *Theoretische Chemie in Deutschland*, in: Nachrichten aus Chemie und Technik 13 (1961), 351–353.

of quantum chemistry Giovanni Battista Bonino[40] published a long paper in the *Gazzetta Chimica Italiana* on the constitution of benzene, in which he acknowledged the importance of Hückel's paper for a fundamental understanding of aromatic compounds.[41] Bonino summarized Hückel's quantum-mechanical theory of aromaticity in the idiom of chemists, enthusiastically welcoming it with the words: "Hückel's theory is certainly the most important physical and mathematical effort in facilitating comprehension of the complicated aromatic structure."[42]

Another country to take note of Hückel's research was England. Lennard-Jones and particularly his assistant Charles Alfred Coulson[43] applied Hückel's ideas about aromatic compounds by means of the MO method to other cyclic compounds[44] as well as to free radicals.[45]

In the New World, Hückel's papers were to some extent forgotten, having to wait until after World War II to be rediscovered in America. There then followed a renaissance of Hückel's Molecular Orbital Theory of π electronic systems. The rising importance in organic chemistry of molecular spectroscopy in conjunction with computer technology were the cause. Problems that had hitherto demanded prohibitive calculational work had come within the range of feasibility in quantum chemistry.[46]

---

[40]Bonino's contribution to the development of quantum chemistry as a separate discipline within the social and political contexts of fascist Italy is discussed in Karachalios, A.: *Giovanni Battista Bonino and the Making of Quantum Chemistry in Italy in the 1930s*, in: Chemical Sciences in the 20th Century, edited by Carsten Reinhardt. Wiley-VCH, Weinheim, 2001, 75–104; *I chimici di fronte al fascismo. Il caso di Giovanni Battista Bonino (1899–1985)*. Istituto Gramsci Siciliano, Palermo, 2001.

[41]Bonino, G. B.: *Sulla costituzione molecolare del Benzolo*, in: Gazzetta Chimica Italiana 65 (1935), 371–423.

[42]"La Teoria di Hückel costituisce certamente lo sforzo fisico-matematico più notevole che mai sia stato fatto per penetrare la difficile struttura del nucleo aromatico." Ibid., p. 399.

[43]Coulson's career in science and his original contribution to the MO method is discussed in Simoes, A., Gavroglu, K.: *Quantum Chemistry qua Applied Mathematics. The Contributions of Charles Alfred Coulson (1910–1974)*, in: Historical Studies in the Physical Sciences 29 (1999), 363–406; Park, B. S.: *Computations and Interpretations: The Growth of Quantum Chemistry*, 1927–1967 (1999), 285 ff.

[44]SBPK, Papers of Erich Hückel, Box 6, Folder 5. 111, Letter Lennard-Jones to Hückel, Cambridge, 22nd April, 1937.

[45]Coulson, C. A.: *The Electronic Structure of Some Polyenes and Aromatic Molecules IV. The nature of the Links of Certain Free Radicals*, in: Proceedings of the Royal Society of London 164 (1938), 383–396; *The Electronic Structur of Some Polyenes and Aromatic Molecules VII. Bonds of Fractional Order by the Molecular Orbital Method*, in: Proceedings of the Royal Society of London 169 (1939), 413–428.

[46]Cf. Brush, S. G.: *Dynamics of Theory Change in Chemistry. Part II: The Benzene Problem 1945–1980*, in: Studies in the Histrory and Philosophy of Science, 30 (1999), 263–302; Streitwieser, A.: *A Lifetime of Synergy with Theory and Experiment*, Jeffrey I. Seeman, Series Editor, American Chemical Society, 1997; *History of Computational Chemistry: A Personal View*, in: Encyclopedia of Computational Chemistry, vol. 2, pp. 1237–1243.

Chapter 6
Hückel's Professional Career in National Socialist Germany

When the National Socialists took over the government in early 1933, Hückel was a lecturer of "chemical physics" at the Württemberg polytechnic (*Technische Hochschule*) in Stuttgart, at the peak of his productivity as a researcher and a teacher. On May 8, 1933 he submitted the fourth and final part of his *opus magnum* on the quantum theory of unsaturated and aromatic compounds to the editorial offices of *Zeitschrift für Physik*.[1] He was planning to offer a number of different courses that academic year, from April 1, 1933 to March 31, 1934.[2] In addition he intended to continue the "Seminar on atomic physics," that he had been offering annually since the academic year 1931/1932 together with the full professor of theoretical physics Peter Paul Ewald.[3]

Ewald had been serving as president of the polytechnic since January 30, 1933, but the fact that he had a Jewish wife eventually made this office untenable. After succeeding in preventing some attacks by National Socialist students, he felt compelled to resign his presidency in protest.[4] Because Ewald himself was an "Aryan" he did not fall under the provisions of the Law for the Restoration of the Professional Civil Service of April 7, 1933 so he was able to retain his professorship until 1937. This law legitimized the first stage of dismissals of Jewish and politically "undesirable" civil servants. Life in Germany under National Socialism became

---

[1] Hückel, E.: *Die freien Radikale der organischen Chemie. Quantentheoretische Beiträge zum Problem der aromatischen und ungesättigten Verbindungen. IV.*, in: ZP 83 (1933), 632–668.

[2] Hückel advertised the following courses in the course catalog of the Württemberg polytechnic in Stuttgart: Theoretical physics II: Thermodynamics, The periodic system in light of atomic theory, Kinetic theory of gases, and Recent conceptions on the nature of solutions. See *Programm der Württembergischen Technischen Hochschule Stuttgart für das Studienjahr 1933/34*, Stuttgart.

[3] Ibid.

[4] Cf. Voigt, J. H.: *Die Universität Stuttgart. Phasen ihrer Geschichte*. Wittwer, Stuttgart, 1981; *Hitlers Machtantritt und die TH Stuttgart. Ein geschichtlicher Rückblick in fünf Folgen*, in: Stuttgarter Uni-Kurier. Zeitung der Universität Stuttgart v. Februar, April, Juni, Dezember 1983, Februar 1984; Heiber, H.: *Universität unterm Hakenkreuz. Teil II, Die Kapitulation der Hohen Schulen. Das Jahr 1933 und seine Themen*. Bd. 1 und 2. K. G. Saur, München; London; New York; Paris 1992.

A. Karachalios, *Erich Hückel (1896–1980)*, Boston Studies in the Philosophy of Science 283, DOI 10.1007/978-90-481-3560-8_6,

increasingly intolerable for his Jewish wife and their four children. 1937 he emigrated, initially on his own, to England, moving on a year later to America, from where he sent for his family.[5]

The fragmentary remnants of the course catalogs of the Stuttgart polytechnic covering this period indicate that Ewald continued to offer the joint seminar with Hückel until the academic year 1934/1935.[6]

6.1 Membership in the National Socialist Welfare and German Workers Party

Initially, Hückel did his utmost to distance himself from the new regime. He recounts in his autobiography that he did not know how best to behave so as not to jeopardize his already precarious position at the polytechnic. His college friend Schumm evidently urged him repeatedly to join the party to secure his future career.[7] In March 1934 Hückel decided to join the National Socialist welfare organization (*Volkswohlfahrt* NSV).[8] Hückel's retrospective justification for this decision follows a pattern common among such defensive a posteriori moral arguments. "Since I saw no other way to save my family and also my science, I reluctantly decided to enter the NSV. [. . .] Besides, I considered the NSV harmless enough, because it was, so I thought, at least not the place where anything bad could happen."[9]

The NSV was a nationwide organization of the Nazi government intended to provide aid for underprivileged "fellow members of the party" in economic or social distress. At the same time, it played a pivotal role in implementing Nazi ideology, serving also as a reliable mouthpiece for party propaganda. Its name is closely associated with the charity collection boxes and "hot-pot Sundays" organized particularly by the "winter relief drive" (*Winterhilfswerk*). Hitler had established this drive in the summer of 1933 while the NSV was being founded in the *Reich*. Party leaders thought this winter aid could effectively redress the dire economic and social need in the nation and forestall any danger of political unrest. The most famous campaign, the "winter relief drive," enlisted all members of the NSV as "voluntary aids" to solicit contributions from house to house or on a street corner. The NS welfare organization was closely affiliated with the party and hierarchically structured from the top down: Reich – region – district – municipal zone – quarter – block.[10]

---

[5] Eckert, M.: *Die Atomphysiker*, pp. 169–170.

[6] Programm der Technischen Hochschule Stuttgart für das Studienjahr 1934/35 (1. April bis 31. März). Buchdruckerei Karl Scharr, Vaihingen Stuttgart 1934.

[7] Hückel, E.: *Ein Gelehrtenleben*, p. 138.

[8] Ibid., 139

[9] Ibid.

[10] A detailed description of this National Socialist organization is provided by Vorländer, H.: *Die NSV: Darstellung und Dokumentation einer nationalsozialistischen Organisation*. Harald Boldt Verlag, Boppard am Rhein, 1988.

Hückel entered the party's welfare organization as a low-level official, a "block warden" ["Blockwalter"].[11] He described his duties in his autobiography:

> As a [block warden] I had to collect the membership dues and take part in so-called "indoctrination evenings;" sell "attendance certificates" and occasionally take part in "collections" such as for the "winter relief," standing on the street with a collection tin, in other words, to beg. First of all, I found it unpleasant, second of all, it robbed me of my precious working hours. I often resorted to changing 10 or 20 marks into small change at the bank, dropping that in the tin and dropping it off. Sometimes my wife assumed the task of collecting, for instance, when boutonnieres had to be peddled. She had the tin full of change in half an hour, big coins among them. She was, after all, an attractive woman![12]

Besides costing him valuable time, his membership in the party welfare organization got him into needless trouble that doubtlessly caused him some worry. Once Hückel was supposed to appear as block warden early on a Sunday morning before Christmas to help distribute "presents to the needy."[13] Because he worked until late in the evenings and for that reason was a "late riser," he arrived about an hour too late. Hückel described the reaction of his superior:

> The municipal zone director scolded me soundly and threatened that if I was so lax, I'd be sure never to get a professorship; he would report me to Berlin, etc. Luckily, Anne had a good friend, a former teacher of hers, who wore the golden party badge and knew Hitler personally, because she had lived in his neighborhood. She was thoroughly decent and had not the least idea – as many others, including me – of what would become of the Nazi party. She had a talk with the angry municipal zone director, and she succeeded in pacifying him and making it clear to him that I tended to use my Sundays to work on important scientific research.[14]

It is not clear how much Hückel's ignorance was due to political naiveté and how much to later repression of an unpleasant past. But no changes occurred to his professional position until 1937. That year was a special year for his academic career. When Ewald emigrated, Hückel assumed teaching his lectures on theoretical physics between April 1, 1937 and September 30, 1937.[15] Soon after beginning this teaching assignment on May 1, 1937, Hückel entered the National Socialist Party.[16] Hückel later justified this decision completely on professional grounds: "Without this I would not have received my position at Marburg and would have lost the one at Stuttgart."[17]

---

[11] Staff questionnaire for Hückel Erich, HSB, extraordinary professorship theoretical physics (Prof. Hückel), holding 305a, acc. 1992/55, no. 4299.

[12] Hückel, E.: *Ein Gelehrtenleben*, p. 139.

[13] Ibid.

[14] Ibid., 139–140

[15] Staff questionnaire for Hückel Erich, HSB, extraordinary professorship theoretical physics (Prof. Hückel), holding 305a, acc. 1992/55, no. 4299.

[16] Proof of membership to the NSDAP, HSM, personnel file Hückel, Erich, holding 310, acc. 1992/55, no. 6230.

[17] Hückel, E.: *Ein Gelehrtenleben*, p. 140.

6.2 Extraordinary Professorship in Theoretical Physics at Marburg

On September 1, 1937 Franz Arthur Schulze, full professor of theoretical physics at the Philipps University in Marburg, was scheduled to retire. So with the approval of Schulze and the dean, Eduard Grüneisen, professor of experimental physics there, began to seek opinions from a few fellow theoretical physicists about whom they thought might be a suitable successor. Grüneisen first wrote a letter to Arnold Sommerfeld on April 16, 1937. He initially informed him that a "tenured extraordinary professorship" in theoretical physics would be available and that according to the information he received, there was no prospect "of changing it into a full professorship. So only younger lecturers come into consideration."[18] Grüneisen also mentioned the specific local preferences:

> We Marburgers find it important to have a young researcher at home in modern theoretical physics and productive in this field, who also knows how to interest students in his area. You can imagine that Kneser, Vogt and I urgently would like to have a colleague with whom we could also discuss problems of interest to us in a fruitful way.[19]

Sommerfeld sorted the people he could recommend by age, classifying them as "older" and "younger" candidates. Among the older ones, first place went to Ernst Gerlach Stückelberg von Breidenbach. Sommerfeld thought, "as a Rockefeller fellow in America and Munich," he had "a broad perspective and has a complete command of methods in wave mechanics. Pauli and Wentzel much appreciated his recent research on nuclear physics." The second candidate he mentioned was Erich Hückel, for his accomplishments in quantum chemistry[20]: "His achievements concerning electrolytes (together with Debye) are well-known; he is the best man in Germany on the application of methods of wave mechanics to chemical composition (benzene)." Then came Carl-Friedrich von Weizsäcker, "curr. with Debye at Dahlem, is highly regarded everywhere as Heisenberg's best pupil and friend. He has strong philosophical inclinations and holds much promise in physics as well." In fourth place Sommerfeld mentioned his own pupil, a colleague of Hückel at Stuttgart, Helmut Hönl. As Ewald's assistant Hönl offered the "Seminar on atomic physics" together with Hückel in the academic year 1934/1935. In Sommerfeld's opinion, Hönl was a "fine, productive mind, his latest paper on elementary particles is very interesting as well. Supposedly gives good lectures." Sommerfeld ended his list with the physicist Walter Wessel with the remark: "highly educated,

[18] SN, Archiv NL 89, 008, Grüneisen an Sommerfeld, Marburg, April 16, 1937.

[19] Ibid.

[20] This was not the first time that Sommerfeld named Hückel on a list of candidates for a professorship. In 1934 he also ranked Hückel second as successor to Erwin Fues, professor of theoretical physics at the Hannover polytechnic from 1929. Fues had just accepted a call to the university and polytechnic in Breslau. In arguing in favor of Hückel's candidacy, Sommerfeld pointed out the "fine research on benzene" and the "great achievements in the theory of electrolytes" he had made together with Debye. SN, Archiv NL 89, 030, folder Gutachten, Arnold Sommerfeld to Georg Prange, Mnich, October 29, 1934.

works on fundamental problems, without however having contributed fruitfully to developments thus far. Is supposedly a useful advisor in the laboratory."[21]

Grüneisen then also consulted Heisenberg, who informed Hückel about their conversations:

> This summer Grüneisen already asked me whom I would suggest for the professorship in theor. physics at Marburg. We talked in detail about you, Hönl, von Weizsäcker, and Stückelberg. In view of the specific duties of the position (collaboration with Grüneisen on his problems, collaboration with the chemists), the given sequence of candidates materialized – through no particular influence of mine. [...] So I hope foremost that you will come to Leipzig, but if not there, then Marburg.[22]

Heisenberg's letter thus reveals that Hückel also had chances at a position in Leipzig.

Prior to 1933, the appointment procedure usually was as follows: when a professorship became vacant, the faculty presented to the local minister of culture a list of three candidates. In most cases, the minister then issued an appointment to the name appearing first on the list. With the introduction of the "*Führer* principle" in the fall of 1933, the faculty's nomination rights were transferred to the university president. Involvement by the faculties was reduced to mere advisory status. They passed their appointment proposals on to the president, who then sent it to the local ministry of culture after appending his own opinion. Since the creation of the new Reich Ministry of Science, Education and Culture in May 1934, the final decision was made by Minister Rust. The Reich Ministry required in addition to academic certifications a submission by the head of the University Lecturers League (*Dozentenschaft*) on the political leanings of the nominees. Only then was the finalist officially called by the Reich minister of education after direct consultations with the university president and the regional leader of the University Lecturer's League, who was directly answerable to the university president.[23]

---

[21] SN, Archiv NL 89, 002, Sommerfeld to Grüneisen, Munich, April 29, 1937. The "younger gentlemen" that Sommerfeld took into consideration were listed in alphabetical order:

"Jensen, Hamburg. Did some good research on crystalline ions as Lenz's assistant.
Kohler, Berlin, pupil of von Laue. Versed in crystalline symmetry.
Maus, Munich. Good teacher, generally educated and reliable, also co-published with me.

Meixner, Giessen. Good mathematician and teacher. Research on wave mechanics together with Bechert.
Möglich, Berlin. Collaborated with von Laue but in recently years has supposedly been more active in politics than science.
Sexl, Vienna with Thirring. Has done good work on nuclear physics and hydrodynamics."

[22] SBPK, Papers of Hückel, Box 6, Folder 5.16, Letter Heisenberg to Hückel, Baveno 16.10.37

[23] This summary of the new regulations is basically a shortened version of the relevant sections in Deichmann, U.: *Biologen unter Hitler, Porträt einer Wissenschaft im NS-Staat*. Fischer, Frankfurt am Main, 1995; *Flüchten, Mitmachen, Vergessen*. Wiley-VCH, Weinheim, 2001. The structural changes at German universities upon the introduction of the "Führer principle" are described in greater detail in Hartshorne, E. Y.: *The German Universities and National Socialism*. Unwin Brothers Ltd., London, 1937; Heiber, H.: *Universität unter dem Hakenkreuz*, K.G.Saur, München, 1992.

In the case at hand, the dean of the University of Marburg suggested, on behalf of the Faculty of Philosophy, "the following personalities on the basis of their scientific merit:" First, Friedrich Hund, already second full professor of theoretical physics at Leipzig. Next on the list were mentioned "in alphabetical order": Hönl; Hückel, an untenured extraordinary professor at Stuttgart since April 1935; and Wessel. If Hund was to be appointed, the Faculty presumed that "he would be conferred the title, rights and honors of a full ordinary professorship like the current occupant of the chair, and that the salary be as generous as possible." The dean added about the list of candidates that the political evaluations of Hund and Hönl had been received only after the scientific opinions had arrived, while those on Hückel and Wessel were still pending. Concerning the political evaluation of Hund, he was "not thoroughly satisfied" because the opinion by the Leipzig Lecturers League leader did not coincide with the positive evaluation by the dean at Leipzig. Nevertheless, the dean at Marburg "refrained from" eliminating Hund's candidacy. The dean's comment about Hönl was the same, because in his case "the verdict by the Lecturers League" was "not so irreproachable, as I personally have learned in Stuttgart."[24]

The university president at Marburg agreed with the dean in placing Hund first on the list of nominees and supported the Faculty's proposal as follows:

> The reservations against Mr. Hund brought forward by the leader of the Leipzig Lecturers League are basically to the effect that Mr. Hund is not sufficiently politically active. I do not consider these reservations compelling, however. It is frequently the case that prominent scholars entirely positively disposed to the National Socialist state do not engage themselves politically, because they are simply not made for such activities. This seems to be the case with Mr. Hund. In view of his generally acknowledged prominent importance in science, I would very much welcome it if Mr. Hund received the call to Marburg.[25]

The president's remarks about the second-place candidates were as follows: "On the basis [of Hönl's] scientific achievements, it would appear justifiable to place him in second place together with Mr. Hückel."[26] The dean approved nomination of Wessel with the following qualification: "Should there be any political reservations about him, I would have to request that we not consider calling him."[27]

It is clear that the Faculty was pushing for Hund's candidacy through the president and the dean by virtue of his scientific prominence. The choice between Hönl and Hückel, as scientifically equivalent alternatives, was left to the Reich Ministry of Science, Education and Culture. It was sure to reach its decision on political grounds rather than on the basis of scientific evaluations.

---

[24] Der Dekan der Philosophischen Fakultät der Philipps-Universität an den Herrn Reichs- und Preussischen Minister für Wissenschaft, Erziehung und Volksbildung, Marburg, 9. Juli 1937, Nr. 402/881, HSM, Extraordinariat Theoretische Physik (Prof. Hückel), Bestand 305a, acc. 1992/55, Nr. 4299.

[25] Der Rektor der Philipps-Universität II A 17/17, Marburg, den 26. Juli 1937, HSM, Exstraordinariat Theoretische Physik (Prof. Hückel), Bestand 305a, acc. 1992/55, Nr. 4299.

[26] Ibid.

[27] Ibid.

Irrespective of Hönl's "not so irreproachable" political evaluation, the fact that he was a pupil of Sommerfeld[28] was reason enough for those responsible at the Reich Ministry for him to fall into as much disfavor as his teacher. Given this situation, we can assume that, compared to Hund and even Hückel, Hönl had little chance of receiving the appointment to Marburg, despite his otherwise unobjectionable political record.

In the fall of 1937 the first official announcement in Hückel's favor arrived from Berlin. By decree of the Reich Minister of Science, Education and Culture dated October 29, 1937, Hückel was appointed with "substituting for the vacated professorship in theoretical physics in the Faculty of Philosophy at the University of Marburg."[29]

Two months later Hückel received the following invitation from Senior Government Official Dames the responsible expert at the Reich Ministry: "I intend to suggest to my Minister here your appointment to the vacated chair in theoretical physics at the University of Marburg and invite you to appear at my Ministry on Saturday, the 11th of December 1937 [...] concerning the closure of the appointment contract."[30]

Hückel describes his meeting with the expert at the Reich Ministry in Berlin in his autobiography as follows:

> The expert official there by the name of "Dames" was impolite, producing the birth certificates and other personal certificates of mine and of my ancestors in order to ascertain whether it wasn't possible to find at least one dark Jewish blemish on my white "Aryan waistcoat". [...]
>
> When after hours of rifling through my papers [...] Mr. Dames had found nothing, I hoped finally to be able to pose my questions concerning the professorship. I asked him to consider that theoretical physics today really was a subject of some major scientific importance, so it really ought to be represented at a university by a full professorship, not an extraordinary position that requires an assistant to be able to meet its obligations in teaching and research. "Friendly" Mr. Dames waved it all curtly aside. The extraordinary professorship was a budgeted position for one extraordinary post; he had no need for an assistant and a secretary. If I did not accept the professorship in the way it was offered to me, I would not get it: it was anyway questionable whether I was politically reliable enough.[31]

Hückel accepted the conditions imposed by the Ministry and was then appointed on April 2, 1938 to the lifetime position as extraordinary professor of theoretical

---

[28]Gutachten über Dozent Dr. Helmut Hönl – Stuttgart, Marburg 9. Juli 1937, HSM, Personalakten Hückel, Erich, Bestand 310, acc. 1992/55, Nr. 6230: "Dr. Helmut Hönl, born Feb. 10, 1903, is a pupil of Arnold Sommerfeld – Munich."

[29]Der Reichs- und Preußische Minister für Wissenschaft, Erziehung und Volksbildung, WP 2687, an den Herrn Universitätskurator in Marburg, Berlin, 29. Oktober 1937, HSM, Personalakten Hückel, Erich, Bestand 310, acc. 1992/55, Nr. 6230.

[30]Der Reichs- und Preußische Minister für Wissenschaft, Erziehung und Volksbildung, WP 2687 II, an den Herrn Universitätskurator in Marburg, Berlin, 4.12.1937, HSM, Personalakten Hückel, Erich, Bestand 310, acc. 1992/55, Nr. 6230.

[31]Hückel, E.: *Ein Gelehrtenleben*, p. 145.

physics at the University of Marburg.[32] These proceedings show that Hückel may well have owed his appointment to the political reservations against the other two candidates on the list submitted by the nominating committee at Marburg.

Until the end of World War II, Hückel was the only teacher of theoretical physics at the university. So his lectures had to cover every aspect of physics. He was also responsible for the practical laboratory sessions. "It was very stressful. These lectures and practical sessions all had to be thoroughly planned out in advance and well prepared," Hückel later commented.[33] Even during his Stuttgart days Hückel's lectures and laboratory sessions were always very carefully prepared. The president of the Stuttgart polytechnic confirmed this in a letter to his colleague in Marburg, informing him of his qualities as a teacher: "About his pedagogical abilities I can only say that Professor Hückel is an excellent teacher."[34]

Very little time was left for his own research. There was not even enough time, according to Hückel, to make new acquaintances among his fellow colleagues.[35] This was also the time when he starting building his house in Marburg. "Unfortunately, the heavy burden from my work, the after-effects of the stress during the '7 years of shame,' and now the building of the house proved too much for my nerves," Hückel remembered about that period of his life.[36] On December 10, 1938 he had "a fit of nerves," necessitating "sick leave until the end of that winter semester."[37] One can say that, from that date onward, Hückel's life took a turn for the worse, both on the personal and private level as well as on the scientific level.

1938 and during the first months of 1939, the National Socialist state was enjoying the height of its success. The balance of power on the European continent tipped perceptibly in favor of the Third Reich. On September 1, 1939 Germany's attack on Poland triggered World War II. "The calamity played itself out," as Hückel noted in his autobiography.[38] He received "a call-up order to Ebsdorf, a small town near Marburg, as gunner at the air force," because during the First World War he had been a seaplane sailor.[39] From Ebsdorf he received a transfer order, taking a bus to the airport near Giessen. Hückel did not stay there long, because it so happened

[32]Der Reichs- und Preußische Minister für Wissenschaft, Erziehung und Volksbildung, an den Herrn Universitätskurator in Marburg, Berlin, 13. April 1938, Urkunde (Abschrift) WP 1036 (b), HSM, Personalakten Hückel, Erich, Bestand 310, acc. 1992/55, Nr. 6230.

[33]Hückel, E.: *Ein Gelehrtenleben*, p. 146.

[34]Der Rektor der Technischen Hochschule an Se. Magnifizenz den Rektor der Universität Marburg/Lahn, Nr. 2265, Stuttgart, 22. Juli 1937, HSM, Extraordinariat Theoretische Physik (prof. Hückel), Bestand 305a, acc. 1992/55, Nr. 4299.

[35]Hückel, E.: *Ein Gelehrtenleben*, p. 146.

[36]Ibid.

[37]An den Herrn Reichsminister für Wissenschaft, Erziehung und Volksbildung, Betrifft: Erkrankung und Beurlaubung des beamteten außerordentlichen Professors der theoretischen Physik Dr. Hückel, Nr. 7424, HSM, Personalakten, Hückel, Erich, Bestand 310, acc. 1992/55, Nr. 6230.

[38]Hückel, E.: *Ein Gelehrtenleben*, p. 148.

[39]Ibid.

that two soldiers too many had been called up.[40] So Hückel returned to his teaching duties at Marburg.

As would be expected, the university halls were empty and research had largely come to a standstill. Very few students came to listen to Hückel's lectures. But they cost Hückel "the same amount of effort, because preparation and presentation do not change with a diminution in attendance."[41] He even had to add the physics practical session for medical students to his schedule, because the assistants in charge were on the battle fields. He made little progress as a scholar. By the end of the war Hückel had published a single paper that did not even have anything to do with quantum chemistry.[42] It provided a critical survey of the natural optical activity of gases and liquids.

[40] Ibid., 149

[41] Ibid., 150

[42] Hückel, E.: *Zur Theorie der natürlichen optischen Aktivität gasförmiger und flüssiger isotroper Stoffe*, in: Z. Elektrochem. 45 (1944), 13–34.

Chapter 7
The Postwar Years

A few months after the capitulation of the German Reich on May 9, 1945, Hückel's newly built house, that had withstood the ravages of war, was one of many in Marburg to be confiscated by the Americans. Hückel's nervous condition deteriorated again under the heavy psychological pressure. His allusion to this in his autobiography was down to earth:

> I had a nervous breakdown; physically half-starved, psychologically stressed to the extreme, I tried to sedate myself with Luminal. But it had the opposite effect. Many medications have a different, unpredictable effect on me than they should. I began to rave and had to be brought to the mental clinic.[1]

When Hückel was released from the clinic, the denazification tribunals started in the fall of 1945, causing him considerable trouble. Following an ordinance by the American military government of Hessen-Nassau from October 6, 1945, the university president dismissed him from his position on October 11, 1945 "with immediate effect," on the grounds of party membership (1937) and holding the office of block warden in the NSV from 1934–37.[2] As a consequence, his bank accounts were frozen, allowing him to withdraw only 300 reichmarks per month. The Hückel family was plunged into financial difficulty. To try to make ends meet Hückel offered lectures privately, asking for voluntary attendance fees from the participants. "I must have been quite well liked among my students as a teacher as well as a person, because quite a group of them took part in my courses," Hückel gratefully reminisced.[3]

In a letter to Debye from June 22, 1946, Hückel described how he and his family were living and his hopes of being reappointed by the University of Marburg. Debye had emigrated to America in 1940 and was already an American citizen:

> I would like to make use of the long-awaited possibility of writing abroad to give you a brief report about myself. [...] I am still here in Marburg but am at the moment suspended from my office because upon my appointment to Marburg I was pressured to enter the

[1] Hückel, E.: *Ein Gelehrtenleben*, p. 161.

[2] Der Rektor der Universität, Tgb. Nr. 3415, Marburg, den 11.10.1945, HSM, Personalakten, Hückel, Erich, Bestand 310, acc. 1992/55, Nr. 6230.

[3] Hückel, E.: *Ein Gelehrtenleben*, p. 162.

A. Karachalios, *Erich Hückel (1896–1980)*, Boston Studies in the Philosophy of Science 283, DOI 10.1007/978-90-481-3560-8_7,

party and filled the post of block warden of the NSV in Stuttgart between '34 and '37 in order to safeguard my position there. But since my overall anti-National Socialistic attitude was generally known here, particularly also from a professional point of view, and my reappointment is desired by the Faculty and is recommended by the university officer, I can hope soon to be returned to my post.

With the birth of our baby daughter Irene in 1944 destiny has been gracious to the family, increasing it to four children. My oldest son, whom you may still vaguely remember from Leipzig and who is now almost 18, could fortunately be spared from senseless sacrifice right at the end of the war. He will take his school leaving examinations sometime next year and plans to study physics and chemistry. As he is quite talented, I would very much like to arrange that he first get a thorough education here and later possibly be given the opportunity to work in science in the USA sometime. But these are perhaps impermissible extrapolations into the future.

Our house, which we had built here shortly before the war, fortunately was spared from the bombing. Research has been generally quite hampered by the war situation. But I did not accept any kind of military projects. Most recently I was working on the theories of natural optical activity and published a summary exposition on it in the Z[eitschrift für] Elektrochim[ie]. Now research is being severely impeded by the overall living conditions, particularly the extreme inability to concentrate that we all have to complain about as a consequence of bad nutrition. Since about 1941 foreign journals have been for the most part inaccessible to us.[4]

Almost a full month later, Hückel was returned his office "as tenured extraordinary professor of theoretical physics and director of the department of theoretical physics at the Institute of Physics of the University of Marburg" as of August 1, 1946. The decree was issued by the president on July 16, 1946 with the approval of the military government.[5] Hückel's work in the following years was frequently interrupted, however, by serious illness and bouts of depression. Hückel never recovered his creativity of the 1930s. After a few unsuccessful attempts to interpret the spectra of aromatic compounds, Hückel decided to give up.[6] He explained his reasoning behind the decision as follows:

There was no point in trying to continue to work in the scientific field that my research on benzene and my associated analyses had sprung from. It was not possible to follow the developments that this field had undergone particularly in the USA and in England, or even to study them retroactively.[7]

The only time Hückel returned to his theory of unsaturated and aromatic compounds since the end of the war until his retirement in 1962 was on the occasion of the 56th plenary session of the Deutsche Bunsen Gesellschaft, that convened in Kiel from May 30 to June 2, 1957. Its guiding theme was "Theory and experiments on the problem of chemical bonding," and the whole event was organized by Hermann Hartmann, the leading quantum chemist of the postwar period. Hartmann invited

[4]SBPK, Papers of Erich Hückel, Box 6, Folder 5. 23, Letter Hückel to Debye, Marburg, 22 Juni 1946.

[5]Der Rektor der Universität an Herr Prof. Hückel, Tgb. Nr. 4855, Marburg, den 16.07.1946, HSM, Personalakten, Hückel, Erich, Bestand 310, acc. 1992/55, Nr. 6230.

[6]*Chemiker im Gespräch: Erich Hückel*, p. 186.

[7]Hückel, E.: *Ein Gelehrtenleben*, p. 166.

Hückel to speak about the topic[8] and Hückel's talk during that conference was "On the modern theory of unsaturated and aromatic compounds."[9] By his own admission, it took him half a year to prepare for it, because he had to inform himself about "how this field had developed."[10] Hückel basically continued to defend the bold main ideas of his theory from the 1930s and tried to conform it to the current state of research.

In 1961 Hückel was promoted to a full professorship, or "*Ordinarius.*" With a mixture of irony and sarcasm he recounted: "That cost the province of Hessen nothing. 1962, I requested my retirement."[11] It was basically a retrospective acknowledgment of his scientific achievements from the 1920s and particularly the 1930s. The university president had petitioned the Hessian Minister of Education and Culture on behalf of the Faculty of Philosophy of the Philipps University, to appoint Hückel as a "personal *Ordinarius.*" According to the Faculty, it was an "appropriate form of personally honoring a researcher of merit."[12] In support of this petition the Faculty argued:

> For decades Mr. Hückel has counted among the famous theoretical physicists of Germany. There are two groups of research in which he contributed significantly toward the development of his discipline. First, in collaboration with P. Debye at the beginning of the twenties he brought the theory of electrolytes a substantial step forward. [...]
>
> The second group of research papers that Mr. Hückel addressed since 1930 builds upon the theory of the benzene ring. In the foregoing century attempts had already been made to develop a model of the binding states of the benzene molecule from the point of view of chemistry. Yet the model by Kekulé, which had become the standard, remained fundamentally enigmatic – for all its usefulness for practical chemistry. Only quantum mechanics allowed a deeper understanding, at the same time expanding on Kekulé's model with the development of the concept of quantum-mechanical resonance. Mr. Hückel created this quantum-mechanical theory of the benzene ring in numerous publications during the thirties. He thereby not only contributed decisively to our understanding of the structure of aromatic compounds in general, but also counts among the founders of the modern quantum theory of chemical bonding.[13]

In 1965 at the centennial celebration of Kekulé's formula for benzene, Hückel was awarded the Otto Hahn prize for chemistry and physics for his theory on aromatic compounds. In 1966 the Stuttgart polytechnic conferred him the honorary degree Dr. rer. nat. e. h. – "probably as compensation for the '7 years of shame,'" was Hückel's characteristically wry comment. Hückel was awarded a number of other distinctions before his death on February 16, 1980 for his accomplishments in the field of quantum chemistry.

---

[8] Ibid., pp. 166–167.

[9] Hückel, E.: *Zur modernen Theorie ungesättigter und aromatischer Verbindungen*, in: Z. Elektrochem. 61 (1957), 866–890.

[10] Ibid.

[11] Ibid., p. 167.

[12] An der Herrn Hess, Minister für Erziehung und Volksbildung durch Seine Magnifizenz den Herrn Rektor der Philipps – Universität Marburg, Nr. 39/66, Marburg. November 1960, HSM, Personalakten Hückel, Erich, Bestand 310, acc. 1992/55, Nr. 6230.

[13] Ibid.

Chapter 8
Summary and Concluding Remarks

The focus of this study has been Erich Hückel's professional passage from physics to physical chemistry to quantum chemistry. His scientific contributions during the 1930s paved the way for the emergence of quantum chemistry as an independent discipline in Germany, in research as well as in teaching. The critical exposition of his main research program in quantum chemistry reveals a hitherto neglected aspect in the history of science: the importance of Erich Hückel's work on the quantum-theoretical foundations of organic chemistry.

As we have seen, the period from 1927 to 1937 is a particularly interesting time in the development of quantum chemistry. The most creative stage of Erich Hückel's scientific career falls precisely within this period. Family influences played an important role, primarily his father's contagious enthusiasm for science that all three of his sons were able to benefit from. Another important factor in Hückel's upbringing was the close ties to the university town of Göttingen. It was there that Hückel studied physics and mathematics, taking his doctorate under Peter Debye with an experimental thesis in the border zone between solid-state chemistry and molecular physics, on the scattering of Röntgen rays by anisotropic fluids. It was also there, at Hilbert's Institute of Mathematics, that Hückel accepted an auxiliary assistantship in 1921. For a year he was immersed in Hilbert's epistemological considerations and ideas. Hilbert's axiomatic thought left its mark on Hückel's quantum theoretical treatment of benzene and other aromatic compounds a few years later. Another assistantship under Max Born acquainted Hückel with an entirely different conceptual and working approach in science. Their subsequent collaboration in Zurich from 1922 to 1928 resulted in their famous research on the theory of strong electrolytes. Their joint paper marked the end of Hückel's first period of scientific creativity. Hückel's "apprenticeship" with Debye was solid methodological preparation for his pioneering achievements in the emerging boundary discipline of quantum chemistry.

The final years leading to 1930 were Hückel's wanderyears. Stimulating exchanges with leading physicists and chemists led to an important professional turning point. With a research scholarship in hand from the International Education Board, he was welcomed by Frederick Donnan at London, and Paul Dirac at Cambridge. A grant from the Rockefeller Foundation opened the door to Niels

A. Karachalios, *Erich Hückel (1896–1980)*, Boston Studies in the Philosophy of Science 283, DOI 10.1007/978-90-481-3560-8_8,

Bohr's famed institute in Copenhagen. Bohr inspired Hückel's reorientation toward conducting a quantum-mechanical inquiry into double bonding. With a grant by the Notgemeinschaft 1929–1930 he was able to join Werner Heisenberg and Friedrich Hund at Leipzig to investigate this topic further.

The move to Leipzig, although at Debye's instigation, promoted his independence as a researcher. He turned away from this teacher's field of physical chemistry to embark on his own project in quantum chemistry. As lecturer of "chemical physics" at the polytechnic in Stuttgart from October 1930 to November 1937, Hückel continued to pursue this agenda. It essentially constituted his quantum theory of unsaturated and aromatic compounds.

Section 2.1 gave a historical survey of the essential ideas underlying Hückel's quantum theory of double bonding. The subsequent comparative analysis of Hückel's model against those advanced by Pauling and Slater examined the respects in which Hückel's theory overtook van't Hoff's classical model of double bonding. Hückel's American rivals were attempting to prop this stereochemical model up with quantum theory. As I have shown, Hückel could not have taken this step forward without his brother Walter's critical support. The physical considerations of this organic chemist on controversial issues in chemistry, published in the early 1920s, served as a basis for Erich Hückel's quantum mechanical analysis of ethylene. The scientific exchanges between the two brothers arising from their collaboration on Walter's two-volume work on the theoretical foundations of organic chemistry were discussed in detail. Walter's research and theoretical approaches, particularly concerning his treatment of stereochemical issues of cyclic compounds, were sketched along with his influence on his younger brother.

How Erich Hückel arrived at his quantum mechanical theory of aromaticity was then examined from a methodological point of view as well as concerning its content. The number of electrons forming a "closed electron shell" stood at the focus of Hückel's explicative approach toward the assignment of a substance among the class of aromatic compounds, not structural characteristics. Hückel's theory of aromaticity could explain the different chemical behaviors of benzene, cyclobutadiene and cyclooctatetraene, which had hitherto remained an anomaly, according to classical structural theory in organic chemistry. Hückel's theory went beyond explaining the characteristics of aromatic compounds already familiar to organic chemists. It predicted the existence of other aromatic ring compounds and inspired their chemical synthesis. Thus with his theory of aromaticity Hückel contributed importantly toward a methodological and conceptual autonomy for the young discipline of quantum chemistry.

As a vanguard in quantum chemistry, Hückel's theory of unsaturated and aromatic compounds had to fight a bitter battle for acceptance. Like its mother discipline of quantum mechanics, quantum chemistry too was initially repudiated by the majority of chemists as unacceptably unintuitive. Quantum mechanics drew the unsettling conclusion that it was, in principle, impossible to describe atomic processes by means of the more immediately comprehensible concepts of classical mechanics. Atomic processes could not be accurately described as motions of particles or waves in normal space. In the mathematical description of atomic events, space of a higher dimension was necessary, what is referred to as configurational space.

Hückel ran up against this resistance in his attempt to formulate a theoretical interpretation of the constitution of benzene in his paper from 1931. He used qualitative arguments, that is symmetry considerations, to contend that it was not possible to describe the benzene molecule by means of a structure formula with alternating localized double and single bonds. Friedrich Hund's localization conditions also appear in the discussion of Hückel's theoretical model of benzene because they determine when it is permissible to correlate a localized bond between two electrons with a chemist's conventional valence line.

Hückel's paper on the benzene problem was soon followed by Linus Pauling's first paper on the same topic, copublished with his student, George Wheland. This triggered a controversy between Hückel and the two Americans about Pauling's differing physical interpretation of spin-invariant theory or his quantum mechanical theory of resonance. In my detailed discussion I also mention why Hückel preferred to use the expression "special portion of energy" over Pauling's term "resonance" between various valence structures and the associated "resonance energy."

In the mid-1930s Hückel's controversy with Pauling took on a new dimension when Wheland and Pauling proposed their quantum theoretical explanation of substitution reactions on a substituted benzene molecule. In essence they attempted to provide a quantum theoretical basis for Christopher Ingold's electronic theory of organic reactions. They chose to apply, not the "valence bond method" that they had otherwise preferred, but the "molecular orbital method." The reasons for this surprising turn were discussed along with Hückel's response and Pauling's new magnetic criterion for defining aromaticity. Hückel integrated Pauling's basic idea of the cyclic motion of p_z electrons into his own theory of aromatic compounds in an interesting way. Hückel thus arrived at a new physical description of the electronic states of benzene.

Aside from a few rare exceptions, the overwhelming majority of chemists apparently did not understand Hückel's theoretical considerations. Hückel's efforts to promote his theory among chemists and chemical physicists in various talks and survey articles could not change this state of affairs. The reason for this failure lay in the experimental and heuristic mentality predominating among chemists in Germany. The traditional training impeded a chemist's ability to adjust conceptually to an unintuitive and purely mathematical treatment of problems in their field. Moreover, as I have described, Hückel's clumsy communication skills, especially when set against Pauling's persuasive charm, had negative consequences on general acceptance of his pioneering results. Besides these traditionalist tendencies among German chemists, there were other institutional and ideological factors at work during the National Socialist regime that had negative consequences not just on Hückel's theories but also on the further development of quantum chemistry as a whole inside Germany.

Appendix

Abbreviations Used for Frequently Cited Journals

B	Berichte der Deutschen Chemischen Gesellschaft
CB	Chemische Berichte: In Fortsetzung der Berichte der Deutschen Chemischen Gesellschaft
CR	Chemical Reviews
Helv. Chim. Acta	Helvetica Chimica Acta
JACS	Journal of the American Chemical Society
JCP	Journal of Chemical Physics
PZ	Physikalische Zeitschrift
Verh. Deut. Phys. Ges.	Verhandlungen der deutschen physikalischen Gesellschaft
Z. Elektrochem.	Zeitschrift für Elektrochemie und angewandte physikalische Chemie.
ZP	Zeitschrift für Physik
ZPC	Zeitschrift für physikalische Chemie
ZPC (B)	Zeitschrift für physikalische Chemie B

Archives

AHQP	Archive for the History of Quantum Physics (microfiche collection, Deutsches Museum, Munich)
BALM	Bunsen-Archiv im Liebig-Museum, Gießen
DMA	Deutsches Museum München, Archiv
HSM	Hessisches Staatsarchiv Marburg
MPG-Archiv	Archiv zur Geschichte der Max-Planck-Gesellschaft Berlin
NBA	Niels Bohr Archiv Kopenhagen
NSUB	Niedersächsische Staats- und Universitätsbibliothek Göttingen, Abteilung Handschriften und seltene Drucke
NTNU	Norges teknisk-naturvitenskapelige universitet, Trondheim
PA	Personnel file
RAC	Rockefeller Archive Center, North Tarrytown, New York

A. Karachalios, *Erich Hückel (1896–1980)*, Boston Studies in the Philosophy of Science 283, DOI 10.1007/978-90-481-3560-8,

SBPK	Staatsbibliothek zu Berlin – Preußischer Kulturbesitz
SN	Sommerfeld Papers, Deutsches Museum , Munich
UA-BASF	Company Archive of BASF AG, Ludwigshafen am Rhein
UAG	Georg-August-Universität Göttingen, University Archive
UAT	Universitätsarchiv Tübinge

Bibliography

Alexejeff, W.: Über die Bedeutung der symbolischen Invariantentheorie für die Chemie, in: ZPC 36 (1901), 741–743.

Anderson, J. D.: A History of Aerodynamics and Its Impact on Flying Machines. Cambridg University Press, Cambridge, 1997.

Antoniotti, P.: E. Hückel e il suo contributo alla chimica quantistica, in: Franco Calascibetta-Eugenio Torracca (a cura di), Atti del II Convegno Nazionale di Storia e Fondamenti della Chimica, Roma, Accademia Nazionale delle Scienze detta die XL, 1988, 287–295.

Arabatzis, T.: The Electron: A Biographical Sketch of a Theoretical Entity. Princeton University, Ph. D. Thesis, 1995.

Arabatzis, T.: How the electrons spend their leisure time: philosophical reflections on a controversy between chemists and physicists, in: The Autonomy of Chemistry, 3rd Erlenmeyer – Colloquy for the Philosophy of Chemistry, Edited by Peter Janich and Nikolaos Psarros, Königshausen & Neumann, Würzburg 1998, 149–159.

Arabatzis, T.: Representing Electrons: A Biographical Analysis of a Theoretical Entity. The University of Chicago Press, Chicago and London, 2006.

Arabatzis, T., Gavroglu, K.: The chemists' electron, in: European Journal Physics 18 (1997), 150–163.

Armit, J. W., Robinson, R.: Polynuclear heterocyclic aromatic types. Part II. Some anhydronium bases, in: Journal of the Chemical Society (London) 127 (1925), 1604–1618.

Arndt, F. und Eistert, B.: Über den "Resonanz"- und "Zwischenstufen"-Begriff bei organischen Substanzen mit mehrfachen Bindungen und die Elektronenformeln, in: ZPC-B 31 (1935), 125–131.

Arndt, F., Scholtz, E. und Nachtwey, P.: Über Dipyrylene und über die Bindungsverhältnisse in Pyron-ringsystemen, in: B. 57 (1924), 1903–1911.

Atkins, P. W.: Einführung in die Physikalische Chemie. VCH, Weinheim, 1993.

Bamberger, E.: Beiträge zur Theoriesechsgliedriger Ringsysteme, in: Liebig's Annalen der Chemie 257 (1890), 1–55.

Bamberger, E.: Ueber die Constitution fünfgliedriger Ringsysteme, in: B. 24 (1891), 1758–1764.

Bamberger, E.: Zur Frage der Valenz des Pyrrolstickstoffatoms, in: B. 26 (1893), 1946–1947.

Bamberger, E.: Zur Constitution fünfgliedriger Ringsysteme, in: Liebig's Annalen der Chemie 273 (1893), 373–379.

Beneke, K.: Erich Hückel (09.08.1896 Berlin-16.02.1980 Marburg), in: Biographien und wissenschaftliche Lebensläufe von Kolloidwissenschaftlern, deren Lebensdaten mit 1996 in Verbindung stehen. Beiträge zur Geschichte der Kolloidwissenschaften, VIII. Mitteilungen der Kolloid-Gesellschaft, 1999. Verlag Reinhard Knof, Nehmten, 1999.

Berson, J. A.: Erich – Hückel – Pionier der Organischen Quantenchemie: Leben, Wirken und späte Anerkennung, in: Angewandte Chemie 108 (1996), 2923–2937.

Berson, J. A.: Erich Hückel, Pioneer of Organic Quantum Chemistry: Reflections on Theory and Experiment, in: Angewandte Chemie International Edition in English 35 (1996), 2750–2764.

Berson, J. A.: Chemical Creativity. Ideas from the Work of Woodward, Hückel, Meerwein, and Others. Wiley-VCH, Weinheim, 1999.
Bloch, F.: Über die Quantenmechanik der Elektronen in Kristallgittern, in: ZP 52 (1928), 555–600.
Bloch, F.: Zur Theorie des Ferromagnetismus, in: ZP 61 (1930), 206–219.
Bloch, F.: Reminiscences of Heisenberg and the early days of quantum mechanics, in: Physics Today 29 (1976) (12), 23–27.
Blumberg, A., Owens, G.: Energy and Conflict: The Life and Times of Edward Teller. Verso, London, 1976.
Bohr, N.: On the constitution of atoms and molecules. (Part I), in: Philosophical Magazine 26 (1913), 1–25; Part II: Systems containing only a single nucleus, in: Philosophical Magazine 26 (1913), 476–502; Part III: Systems containing several nuclei, in: Philosophical Magazine 26 (1913), 857–875.
Bohr, N.: On the Model of a Triatomic Hydrogen Molecule, in: Medd. Kungl. Vetenskapsakad, Nobelinstitut, 5(28) (1919), 1–16.
Bohr, N.: Über die Serienspektra der Elemente, in: ZP 2 (1920), 423–469.
Bohr, N.: Der Bau der Atome und die physikalischen und chemischen Eigenschaften der Elemente, in: ZP 9 (1922) 1–67.
Born, M.: Über die ultraroten Eigenschwingungen zweiatomiger Kristalle, in: PZ 19 (1918), 539–548.
Born, M. und Landé, A.: Über die Berechnung der Kompressibilität regulärer Kristalle aus der Gittertheorie, in: Verh. Dtsch. Physik Ges. 20 (1918), 210–216.
Born, M.: Über kubische Atommodelle, in: Verh. Dtsch. Physik Ges. 20 (1918), 230–239.
Born, M.: Dynamik der räumlichen Atomstruktur, in: ZP 2 (1920), 83–86.
Born, M.: Bemerkungen über die Grösse der Atome, in: ZP 2 (1920), 87–89.
Born, M.: Würfelatome, periodisches System und Molekülbildung, in: ZP 2 (1920), 380–404.
Born, M. Zur Quantentheorie der chemischen Kräfte, in: ZP 64 (1930), 729–740.
Born, M. und Brody, E.: Über die Schwingungen eines mechanischen Systems mit endlicher Amplitude und ihre Quantelung, in: ZP 6 (1921) 140–152.
Born, M. und Pauli Jr., W.: Über die Quantelung gestörter mechanischer Systeme, in: ZP 10 (1922) 137–158.
Born, M.: Atomtheorie des festes Zustandes (Dynamik der Kristallgitter), in: Encyklopedie der mathematischen Wissenschaften, Band V/3, G. B. Teubner, Leipzig 1923, 527–781.
Born, M.: Chemische Bindung und Quantenmechanik, in: Ergebnisse der Exakten Naturwissenschaften, 10 (1931), 387–444.
Born, M.: Zur Theorie der homöopolaren Valenz bei mehratomigen Molekülen, in: Angewandte Chemie 45 (1932), 6–8.
Born, M.: Welche Vorstellung von der chemischen Bindung vermittelt die Quantenmechanik?, in: Angewandte Chemie 46 (1933), 179–180.
Born, M. und Heisenberg, W.: Die Elektronenbahnen im angeregten Heliumatom, in: ZP 16 (1923), 229–243.
Born, M. und Hückel, E.: Zur Quantentheorie mehratomiger Molekeln, in: PZ 24 (1923), 1–12.
Brock, W. H.: The Norton History of Chemistry. W. W. Norton & Company, New York, 1992.
Brush, S. G.: Dynamics of theory change in chemistry: Part 1. The benzene problem 1865–1945; Part 2. Benzene and molecular orbitals, in: Studies in the History and Philosophy of Science 30 (1999), 21–79, 263–302.
Burk, L. A.: Fritz Arndt and his chemistry books in the Turkish language, in: Bulletin for the History of Chemistry 28 (2003), 42–52.
Burrau, ø.: Berechnung des Energiewertes des Wasserstoffmolekel-Ions ($H^+{}_2$), in: Det Kig. Videnskabernes Selskab Matematisk-Fysiske Meddelelser, 7 (1927), 24–38.
Buttker, K.: Widersprüche der Entwicklung – Entwicklung der Widersprüche. Die Herausbildung der Quantenchemie im Blickfeld philosophischer Analyse. VEB Deutscher Verlag der Wissenschaften, Berlin 1988.

Cassidy, C. D.: Werner Heisenberg: Leben und Werk. Spektrum Akademischer Verlag, Heidelberg, Berlin, Oxford, 1995.

Chayut, M.: From the periphery: the genesis of Eugene P. Wigner's application of group theory to quantum mechanics, in: Foundations of Chemistry 3 (2001), 55–78.

Cerruti, L.: Free Electrons, Lo sviluppo della chimica organica fisica, 1900–1940, in: F. Abbri e M. Ciardi (a cura di), Atti del VIII Convegno Nazionale di Storia e Fondamenti della Chimica (Arezzo, 28–30 Ottobre 1999), Roma, Accedemia Nazionale delle Scienze detta die XL, 1999, 207–263.

Cohen, E.: Jacobus Henricus van't Hoff. Sein Leben und Wirken, Leipzig, 1912.

Coulson, C. A.: The electronic structure of some polyenes and aromatic molecules IV. The nature of the links of certain free radicals, in: Proceedings of the Royal Society of London 164 (1938), 383–396.

Coulson, C. A.: The electronic structure of some polyenes and aromatic molecules VII. Bonds of fractional order by the molecular orbital method, in: Proceedings of the Royal Society of London 169 (1939), 413–428.

Coulson, C. A., O'Leary, B., Mallion, R. B.: Hückel Theory for Organic Chemists. Academic Press, London, 1978.

Davies, M.: Peter J. Debye (1884–1966), in: Biographical Memoirs of Fellows of the Royal Society; The Royal Society, London, vol. 16 (1970), 174–132.

Debye, P. und Scherrer, P.: Interferenzen an regellos orientierten Teilchen im Röntgenlicht. I, in: PZ 17(1916), 277–283.

Debye, P. und Scherrer, P.: Interferenzen an regellos orientierten Teilchen im Röntgenlicht. II, in: PZ 18 (1917), 291–301.

Debye, P. und Hückel, E.: Zur Theorie der Elektrolyte. I. Gefrierpunktserniedrigung und verwandte Erscheinungen, in: PZ 24 (1923), 185–206.

Debye, P. und Hückel, E.: Zur Theorie der Elektrolyte. II. Das Grenzgesetz für die elektrische Leitfähigkeit, in: PZ 24 (1923), 305–325.

Debye, P.: Polare Molekeln. Verlag von S. Hirzel, Leipzig 1929.

Debye, P. (Hrg.): Leipziger Vorträge 1929. Dipolmoment und chemische Struktur. Hirzel Verlag, Leipzig, 1929.

Debye, P.: An Interview, in: Science 145 (1964), 554–559.

Deichmann, U.: Flüchten, Mitmachen, Vergessen: Chemiker und Biochemiker in der NS-Zeit. Wiley-VCH, Weinheim, 2001.

Dewar, M. J. S.: The Molecular Orbital Theory of Organic Molecules. McGraw-Hill, New York, 1969.

Duin, C. F.: Über die künftige Entwicklung der organischen Chemie, in: ZPC 130 (1927), 353–364.

Dumas, J-B.: Ueber die Constitution der Essigsäure und Chloressigsäure, in: Annalen der Chemie und Pharmacie 33 (1840) 179–182.

Dumas, J-B.: Ueber das Gesetz der Substitutionen und die Theorie der Typen, in: Annalen der Chemie und Pharmacie 33 (1840), 259–300.

Ebert, L., Eisenschnitz, R. und Hartel, H.: Über die elektrische Symmetrie des Molekülbaus von Methanabkömmlingen, in: ZPC (B) 1 (1928), 94–114.

Eckert, M.: Die Atomphysiker. Eine Geschichte der theoretischen Physik am Beispiel der Sommerfeldschule. Vieweg, Braunschweig/Wiesbaden, 1993.

Einstein, A. Born, M. :Breifwechsel 1916–1955. Nymphenburger Verlagshandlung, München, 1969.

England, W., Salmon, L. S., Ruedenberg, K.: Localized molecular orbitals: a bridge between chemical intuition and molecular quantum mechanics, in: Fortschritte der chemischen Forschung 23 (1971), 31–122.

Estermann, I.: Dipolmoment und Molekülbau in: Leipziger Vorträge 1929, Dipolmoment und chemische Struktur, Hrg. von P. Debye, Verlag von S. Hirzel, Leipzig, 1929.

Falkenhagen, H. (Hrg.): Leipziger Vorträge 1928. Quantentheorie und Chemie. Hirzel Verlag, Leipzig, 1928.

Falkenhagen, H.: Elektrolyte. Hirzel, Leipzig, 1932.

Francoeur, E.: Molecular models and the articulation of structural constraints in chemistry, in: Tools and Modes of Representation in the Laboratory Sciences, Edited by Ursula Klein. Kluwer Academic Publishers, Dordrecht / Boston / London, 2001.

Frenking, G.: 100. Geburtstag von Erich Hückel, in: Chemie in unserer Zeit 31 (1997), 27–31.

Freudenberg, K. (Hrg.): Stereochemie. Eine Zusammenfassung der Ergebnisse, Grundlagen und Probleme. Franz Deuticke, Leipzig und Wien, 1933.

Furukawa, Y.: Polymer science: from organic chemistry to an interdisciplinary science, in: Chemical Sciences in the 20th Century, Bridging Boundaries, Edited by Carsten Reinhardt. Wiley-VCH, Weinheim, 2001.

Gans, L. und Walden, P.: Zum 70. Geburtstag von A. v. Weinberg, in: Zeitschrift für angewandte Chemie 43 (1930), 703–708.

Garratt, P. und Vollhardt, P.: Aromatizität. Thieme-Verlag, Stuttgart, 1973.

Gavroglu, K.: Fritz London: A Scientific Biography. Cambridge University Press, Cambridge, 1995.

Gavroglu, K.: Appropriating the atom at the end of the 19th century: chemists and physicists at each other's throat, in: Philosophers in the Laboratory, Edited by Valeria Mossini (Academia Nazionale di Scienze, Lettere e Arti di Modena, MUSIS, Roma, 1996, 93–106.

Gavroglu, K.: The physicists' electron and its appropriation by the chemists, in: Histories of the Electron, The Birth of Microphysics, Edited by Jed Z. Buchwald and Andrew Warwick. The MIT Press, MA, 2001, 363–400.

Gavroglu, K., Simões, A.: The Americans, the Germans and the beginnings of quantum chemistry: the confluence of diverging traditions, in: Historical Studies in the Physical and Biological Sciences 25 (1994), 47–110.

Gavroglu, K., Simões, A.: One face or many? The role of textbooks in building the new discipline of quantum chemistry, in: Communicating Chemistry. Textbooks and their Audiences, 1789–1939, Edited by Lundgren, A. and Bensaude-Vincent, Bernardette, Science History Publications, Sagamore Beach, MA, 2000, 415–449.

Gey, E.: Der Zusammenhang von interdisziplinären Forschungssituationen und Kooperationsverhalten bei der Bildung und Entwicklung neuer Spezialgebiete, dargestellt am Beispiel der Quantenchemie, in: Interdisziplinarität in der Forschung: Analysen und Fallstudien, hrsg. von Heirich Parthey und Klaus Scheiber. Akademie-Verlag, Berlin, 1983, 151–175.

Gey, E.: Quantenchemie – eine interdisziplinäre Entwicklumg, in: Wissenschaft und Fortschritt 34 (1984) 11, 282–285.

Gordan, P. und Alexejew, W.: Übereinstimmung der Formeln der Chemie und der Invariantentheorie, in: ZPC 35 (1900), 610–633.

Gortler, L.: The physical organic community in the United States, 1925–50. An emerging network, in: Journal of Chemical Education 62 (1985), 753–757.

Gray, G. W.: Education on an international scale. A history of the International Education Board 1923–1938. Harcourt Brace, New York, 1941.

Haberditzl, W.: 50 Jahre Theorie der chemischen Bindung, in: Zeitschrift für Chemie 18 (1978), 353–359.

Hanack, M.: Das wissenschaftliche Werk von Walter Hückel, in: CB 113 (1980), V-XVII.

Hartmann, H., Longuet-Higgins, H. C.: Erich Hückel, 9 August – 16 February 1980, in: Biographical Memoirs of Fellows of the Royal Society 28 (1982), 153–162.

Hartshorne, E. Y.: The German Universities and National Socialism, Unwin Brothers Ltd., London, 1937.

Hauptmann, S.: Aromatizität–Bedeutungswandel eines Begriffes, in: Zeitschrift für Chemie 13 (1973), 361–364.

Hecht, O.: Otto Schmidt, 8. 9. 1874 – 17. 5. 1943, in: B. 76 (1943), 121–125.

Heiber, H.: Universität unter dem Hakenkreuz. Teil II, Die Kapitulation der Hohen Schulen. Das Jahr 1933 und seine Themen. Bd. 1 und 2. K. G. Saur, München; London; New York; Paris, 1992

Heilbronner, E., Bock, H.: Das HMO-Modell und seine Anwendung. Verlag Chemie, GmbH, Weinheim, 1968.

Heine, J. U.: Verstand & Schicksal – Die Männer der I.G. Farbenindustrie AG in 161 Kurzbiographien, Weinheim, 1990.
Heisenberg, W.: Zur Theorie es Ferromagnetismus, in: Z P 49 (1928), 619–636.
Heitler, W. und London, F.: Wechselwirkung neutraler Atome und homöopolare Bindung nach der Quantenmechanik, in: ZP 44 (1927), 455–472.
Heitler, W.: Zur Quantentheorie der Valenz, in: Naturwissenschaften, 17 (1929), 546–547.
Heitler, W.: Zur Gruppentheorie der homöopolaren chemischen Bindung, in: ZP 47 (1928), 835–858.
Heitler, W.: Der gegenwärtige Stand der quantenmechanischen Theorie der homöopölaren Bindung, in: Physikalische Zeitschrift 31 (1930), 185–204.
Heitler, W. und Rumer, G.: Quantentheorie der chemischen Bindung für mehratomige Moleküle, in: ZP 68 (1931), 12–41.
Heitler, W.: Quantentheorie und homöopolare chemische Bindung, in: Handbuch der Radiologie Band VI, Quantenmechanik der Materie und Strahlung, Teil II Moleküle. Akademische Verlagsgesellschaft M. B. H., Leipzig 1934, 485–586.
Heitler, W.: Quantum chemistry: the early period, in: International Journal of Quantum Chemistry 1 (1967), 13–36.
Hertel, E.: Physikalisch – chemische Problem der Organischen Chemie, in: Z. Elektrochem. 40 (1934), 405–413.
Hiebert, E.: Discipline identification in chemistry and physics, in: Science in Context 9(2) (1996), 93–119.
Hilbert, D.: Natur und mathematisches Erkennen. Herausgegeben und mit einer englischen Einleitung versehen von David E. Rowe. Birkhäuser, Basel, 1992.
Hoddeson, L., Baym, G., Eckert, M.: The development of the quantum-mechanical electron theory of metals, 1928–1933, in: Reviews of Modern Physics 59 (1987), 287–327.
Hoffmann, R. und Laszlo, P.: Darstellungen in der Chemie – die Sprache der Chemiker, in: Angewandte Chemie 103 (1991), 1–16.
Holton, G.: Mach, Einstein, and the search for reality, in: Thematic Origins of Scientific Thought: Kepler to Einstein. Harvard University Press, Cambridge, MA, 1988.
Hückel, A.: Wer ist's Erich Hückel, in: Nachrichten aus Chemie und Technik 13 (1965), 382–383.
Hückel, A.: Die Vaccinekörperchen. Nach Untersuchungen an der geimpften Hornhaut des Kaninchens. Ein experimenteller und kritischer Beitrag zur Frage nach dem Contagium der Vaccine. Gustav Fischer, Jena, 1898.
Hückel, E.: Zerstreuung von Röntgenstrahlen durch anisotrope Flüssigkeiten, in: PZ, 22 (1921), 561–563.
Hückel, E.: Zerstreuung von Röntgenstrahlen durch anisotrope Flüssigkeiten. Dissertation, Göttingen, 1921.
Hückel, E.: Zur Theorie der Elektrolyte, in: Ergebnisse der exakten Naturwissenschaften 3 (1924), 199–276.
Hückel, E.: Zur Theorie konzentrierter wässeriger Lösungen starker Elektrolyte, in: PZ 26 (1925), 93–147.
Hückel, E.: Adsorption und Kapillarkondensation, in: Bd. 7 der Kolloidforschung in Einzeldarstellungen. Akademische Verlagsgesellschaft, Leipzig, 1928.
Hückel, E.: Theorie der Beweglichkeit des Wasserstoff- und Hydroxylions in wässeriger Lösung, in: Z. Elektrochem. 34 (1928), 546–566.
Hückel, E.: Zur Quantentheorie der Doppelbindung, in: ZP 60 (1930), 423–456.
Hückel, E.: Zur Quantentheorie der Doppelbindung und ihres stereochemischen Verhalten, in: Z. Elektrochem. 36 (1930), 641–645.
Hückel, E.: Quantentheoretische Beiträge zum Benzolproblem. I. Die Elektronenkonfiguration des Benzols und verwandter Verbindungen, in: Zeit. Physik 70 (1931), 204–286.
Hückel, E.: Quantentheoretische Beiträge zum Benzolproblem. II. Quantentheorie der induzierten Polaritäten, in: ZP 72 (1931), 310–337.
Hückel, E., Hückel, W.: Theory of induced polarities in benzene, in: Nature 129 (1932), 937–938.

Hückel, E.: Quantentheoretische Beiträge zum Problem der aromatischen und ungesättigten Verbindungen. III., in: ZP 76 (1932), 628–654.

Hückel, E.: Die freien Radikale der organischen Chemie. Quatentheoretische Beiträge zum Problem der aromatischen und ungesättigten Verbindungen. IV., in: ZP 83 (1933), 632–668.

Hückel, E.: Aromatic and Unsaturated Molecules: Contributions to the Problem of their Constitution and Properties, in: International Conference on Physics, Paper and Discussions (vol. II), The Physical Society, London, 1935. The University Press, Cambridge, 1935, 9–35.

Hückel, E.: Composés aromatiques et non saturés. Recherches théoriques concernant leur constitution et leurs propriétés, in: Le Journal de Physique et le Radium 6 (1935), 347–358.

Hückel, E.: Die Bedeutung der neuen Quantentheorie für die Chemie, in: Z. Elektrochem. 42 (1936), 657–662.

Hückel, E.: Kritische Betrachtungen zur Theorie der Substitutionsreaktionen an substituierten Benzolen, in: ZPC (B) 35 (1937), 163–192.

Hückel, E.: Grundzüge der Theorie ungesättigter und aromatischer Verbindungen, in: Z. Elektrochem. 43 (1937), 752–788, 827–849.

Hückel, E.: Zur Theorie der natürlichen optischen Aktivität gasförmiger und flüssiger isotroper Stoffe, in: Z. Elektrochem. 45 (1944), 13–34.

Hückel, E.: Zur modernen Theorie ungesättigter und aromatischer Verbindungen, in: Z. Elektrochem. 61 (1957), 866–890.

Hückel, E.: Erinnerungen an Peter Debye und an meine Lehrjahre, in: Physikalische Blätter 28 (1972), 53–57.

Hückel, E.: Ein Gelehrtenleben. Ernst und Satire. Verlag Chemie, Weinheim, 1975.

Hückel, W.: Beiträge zur Auffassung der Homöopolaren Atombildung, in: Z. Elekrochem. 27 (1921), 305–309.

Hückel, W.: Zur Stereochemie bicyclischer Ringsysteme I. – Die Stereoisomerie des Dekahydronaphthalins und seiner Derivate, in: Liebig‘s Annalen der Chemie 441 (1925), 1–48.

Hückel, W.: Die raumisomeren Formen des Dekahydronaphthalins, in: B 58 (1925), 1449 - 1452.

Hückel, W.: Der gegenwärtige Stand der Spannungstheorie, in: Fortschr. Chem. Physik u. physikal. Chem. 19 (1928), 1–36.

Hückel, W.: Molekülbau und Dipolmoment, in: ZPC (B) 2 (1929), 451–457.

Hückel, W.: Theoretische Grundlage der Organischen Chemie. Erste Auflage, Band I, II Akademische Verlagsgesellschaft m. b. H., Leipzig 1931.

Hückel, W.: Die Fortschritte in der organischen Chemie 1929 – 1931, in: Angewandte Chemie 45 (1932), 457–471.

Hund, F.: Zur Deutung einiger Erscheinungen in den Molekelspektren, in: ZP 36 (1926), 657–674.

Hund, F.: Zur Deutung der Molekelspektren. I, II, IV in: ZP 40 (1927), 742–764; 42 (1927), 93–120; 51 (1928), 759–795.

Hund, F.: Chemical binding, in: Transactions of the Faraday Society 25 (1929), 646–648.

Hund, F.: Zur Frage der chemischen Bindung, in: ZP 73 (1931), 1–30.

Hunf, F.: Zur Frage der chemischen Bindung. II. Zum Verständnis der organischen Chemie, in: ZP 73 (1931), 565–577.

Hund, F.: Zur Theorie der schwerflüchtigen nichtleitenden Atomgitter in: ZP 74 (1932), 1–17.

Hund, F.: Allgemeine Quantenmechanik des Atom– und Molekelbaues, in: Handbuch der Physik (zweite Auflage), Band XXIV, Erster Teil: Quantentheorie. Verlag von Julius Springer, Berlin, 1933, 561–693.

Hund, F.: Description of the binding forces in molecules and crystal lattices on quantum theory, in: International Conference on Physics, Paper and Discussions (vol. II), The Physical Society, London, 1935. The University Press, Cambridge 1935, 36–45.

Hund, F.: Molekelmodelle, in: Das Molekül und der Aufbau der Materie, hrsg. von K. W. Wagner, Friedr. Vieweg & Sohn, Braunschweig, 1949.

Hund, F.: Göttingen, Kopenhagen, Leipzig im Rückblick, in: Bopp, F.: Werner Heisenberg und die Physik unserer Zeit. Vieweg, Braunschweig, 1961.

Hund, F.: Frühgeschichte der quantenmechanischen Behandlung der chemischen Bindung, in: Angewandte Chemie 89 (1977), 89–94.

Ingold, C. K.: The principles of aromatic substitution from the standpoint of the electronic theory of valency, in: Recueil des Travaux Chimiques 48 (1929), 797–812.

Ingold, C. K.: Significance of tautomerism and of the reactions of aromatic compounds in the electronic theory of organic reactions, in: Journal of the Chemical Society (London) (1933), 1120–1127.

Ingold, C. K.: Principles of an electronic theory of organic reactions, in: Chemical Reviews 15 (1934), 225–274.

Ingold, C. K.: Mesomerism and tautomerism, in: Nature 133 (1934), 946–947.

Jaenicke, W.: 100 Jahre BUNSEN-Gesellschaft. Steinkopff, Darmstadt, 1994.

Janich, P. und Psarros, N. (Hrsg.): Die Sprache der Chemie. Königshausen & Neumann, Würzburg, 1996.

Janich, Peter und Psarros, N. (Hrsg.): The Autonomy of Chemistry. 3rd Erlenmeyer-Colloquy for the Philosophy of Chemistry, Königshausen & Neumann, Würzburg, 1996.

Jensen, C.: Two one-electron anomalies in the old quantum theory, in: Historical Studies in the Physical Sciences 15 (1984) 81–106.

Jost, W.: The first 45 years of physical chemistry in Germany, in: Annual Review of Physical Chemistry 17 (1966), 1–14.

Karachalios, A.: Die Entstehung und Entwicklung der Quantenchemie in Deutschland, in: Mitteilungen der Gesellschaft Deutscher Chemiker Fachgruppe Geschichte der Chemie 13 (1997), 163–179.

Karachalios, A.: On the making of quantum chemistry in Germany, in: Studies in the History and Philosophy of Modern Physics, 31B (2000), 493–510.

Karachalios, A.: Giovanni Battista Bonino and the making of quantum chemistry in Italy in the 1930s, in: Chemical Sciences in the 20th Century. Bridging Boundaries, Edited by Carsten Reinhardt. Wiley-VCH, Weinheim, 2001, 75–104.

Karachalios, A.: I chimici di fronte al fascismo. Il caso di Giovanni Battista Bonino (1899–1985). Istituto Gramsci Siciliano, Palermo, 2001.

Kauffmann, G. B.: Alfred Werner – Founder of Coordination Chemistry. Springer, Berlin, Heidelberg, New York, 1966.

Kaufmann, G. B.: Foundation of nitrogen stereochemistry, Alfred Werner's inaugural dissertation, in: Journal of Chemical Education 43 (1966), 155–165.

Kaufmann, G. B.: The stereochemistry of trivalent nitrogen compounds: Alfred Werner and the controversy over the structure of oximes, in: Ambix, 19 (1972), 129–144.

Kermack, W. O., Robinson, R.: An explanation of the property of induced polarity of atoms and an interpretation of the theory of partial valencies on an electronic basis, in: Journal Chemical Society (London) 121 (1922), 427–440.

Kikuchi, S.: Aromatic seven-membered ring compounds and quantum chemistry – the first verification of Hückel's rule and 1st influence, in: Kagakusi Kenkyu. Journal of History of Science, Japan Series II 29(174) (1990), 65–73.

Kikuchi, S.: Development of the electronic formula of benzene, in: Kagakusi Kenkyu. Journal of History of Science, Japan Series II 29(176) (1990), 209–218.

Kikuchi, S.: A history of the structural theory of benzene – the aromatic sextet rule and Hückel's rule, in: Journal of Chemical Education 74 (1997), 194–201.

Knoll, P.: Otto Lehmann, Erforscher der flüssigen Kristalle. P. M. Knoll, Ettlingen, 1988.

Kragh, H.: Chemical aspects of Bohr`s 1913 theory, in: Journal of Chemical Education 54 (1977), 208–210.

Kragh, H.: Niels Bohr`s second atomic theory, in: Historical Studies in the Physical and Biological Sciences 10 (1979), 123–186.

Kragh, H.: Historiography of electronic valence theory, in: Annals of Science 40 (1983), 289–295.

Kragh, H.: Bohr's atomic theory and the chemists, 1913–1925, in: Rivista di Storia della Scienza 2 (1985) 463–486.

Kragh, H.: Before quantum chemistry: Erich Hückel and the physics-chemistry interface, in: Centaurus 43 (2001), 1–16.

Kratzer, A.: Die ultraroten Rotationsspektren der Halogenwasserstoffe, in: ZP 3 (1920), 289–307.

Kratzer, A.: Eine spektroskopische Bestätigung der Isotopen des Chlors, in: ZP 3 (1920), 460–465.

Krätz, O.: Das Portrait: Jacobus Henricus van't Hoff 1852–1911, in: Chemie in unserer Zeit 8 (1974), 135–142.

Kurt, H.: Stereochemie vor van't Hoff und Le Bel, in: Chemie in unserer Zeit 8 (1974), 129–134.

Kutzelnigg, W.: Einführung in die Theoretische Chemie. Band 2: Die chemische Bindung. VCH, Weinheim, New York, 1978.

Kutzelnigg, W.: Friedrich Hund und die Chemie, in: Angewandte Chemie 108 (1996), 629–643.

Landé, A.: Dynamik der räumlichen Atomstruktur, in: Verh. Dtsch. Physik Ges. 21 (1919), 2–12, 644–662.

Lapworth, A.: A theoretical derivation of the principle of induced alternate polarities, in: Journal of the Chemical Society (London) 121 (1922), 416–427.

Lapworth, A., Robinson, R.: Distribution of electrons in the aromatic nucleus and the early stages of aromatic substitutions, in: Nature 129 (1932), 278.

Lapworth, A., Robinson, R.: Theory of induced polarities in benzene, in: Nature 130 (1932), 273.

Lehmann, O.: Flüssige Kristalle, sowie Plastizität von Kristallen im Allgemeinen, Molekulare Umlagerungen und Aggregatzustandsänderungen. Engelmann-Verlag, Leipzig, 1904.

Lennard-Jones, J. E.: The electronic structure of some diatomic molecules, in: Transactions of the Faraday Society 25 (1929), 668–686.

Londale, K.: Magnetic anisotropy and electronic structure of aromatic molecules, in: Proceedings of the Royal Society of London 159 (1937), 149–161.

London, F.: Zur Quantentheorie der homöopolaren Valenzzahlen, in: ZP 46 (1928) 455–477.

London, F.: Zur Quantenmechanik der homöopolaren Valenzchemie, in: ZP 50 (1928), 24–51.

London, F.: Théorie quantique du diamagnétisme des combinaisons aromatiques, in: Comptes Rendus 205 (1937), 28–30.

London, F.: Supraconductivity in aromatic molecules, in: JCP 5 (1937), 837–838.

London, F.: Théorie quantique des courants interatomiques dans les combinaisons aromatiques, in: Journal de Physique et le Radium 8 (1937), 397–409.

Maier, G.: "Aromatisch"–was heißt das eigentlich?, in: Chemie in unser Zeit 9 (1975), 131–141.

Mark, H.: Die Ermittlung von Molekülstrukturen durch Beugung von Elektronen an einen Dampfstrahl, in: Z. Elektrochem. 36 (1930), 675–676.

Mehra, J., Rechenberg, H.: *The Historical Development of Quantum Theory*. Springer-Verlag, New York, Heidelberg, Berlin, 1982.

Meinel, C.: Die Chemie an der Universität Marburg seit Beginn des 19. Jahrhunderts: Ein Beitrag zu ihrer Entwicklung als Hochschulfach. Academia Marburgensis, hrsg. Von der Philipps – Universität Marburg, Bd. 3, Elwert, Marburg, 1978.

Meinel, C.: Freudenberg, Karl Johann, in: Dictionary of Scientific Biography, Supplement II, hrsg. Von Frederick L. Holmes, Ch. Scribner's Sons, New York, 1990.

Minkin, V. I., Glukhovtsev, M. N., Simkin, B. Y. A.: Aromaticity and Antiaromaticity. New York, 1994.

Mohr, E.: Die Baeyersche Spannungstheorie und die Struktur des Diamanten, in: Journal für Praktische Chemie (Neue Folge) [2] 98 (1918), 315–353.

Mohr, E: Zur Theorie der cis-trans-Isomerie des Dekahydro-naphtalins, in: B 55 (1922) 230–231.

Morawetz, H.: Herman Mark, life and accomplishments, in: Macromol. Symp. 98 (1995), 1173–1184.

Morris, P. J. T., Travis, A. S., Reinhardt, C.: Research fields and boundaries in twentieth-century organic chemistry, in: Chemical Sciences in the 20th Century. Bridging Boundaries, Edited By Carten Reinhardt. Wiley-VCH, Weinheim, 2001.

Mosini, V.: A brief history of the theory of resonance and its interpretation, in: Studies in History and Philosophy of Modern Physics 31B (4) 2000, 569–581.

Mulliken, R. S.: The assignment of quantum numbers for electrons in molecules. I, in: Physical Review 32 (1928), 186–222; The assignment of quantum numbers for electrons in molecules.

II. Correlation of molecular and atomic electron states, in: Physical Review 32 (1928), 761–772; The assignment of quantum numbers for electrons in molecules. III. Diatomic hydrides, in: Physical Review 33 (1929), 730–747.
Mulliken, R. S.: Elektronenzustände und chemische Bindung in zweiatomigen Molekülen, in: Z. Elektrochem. 36 (1930), 603–605.
Mulliken, R. S.: Bonding power of electrons and theory of valence, in: Chemical Reviews, 9 (1931), 347–388.
Mulliken, R. S.: The interpretation of band spectra. I, II, III, in: Reviews of Modern Physics 2 (1930), 60–115; 3 (1931), 89–155; 4 (1932), 1–86.
Mulliken, R. S.: Electronic structures of polyatomic molecules and valence. I, in: Physical Review 40 (1932), 55–62.
Mulliken, R. S.: Electronic structures of polyatomic molecules and valence. II. General Considerations, in: Physical Review 41 (1932), 49–71.
Mulliken, R. S.: Electronic structures of polyatomic molecules and valence. III. Quantum theory of the double bond, in: Physical Review 41 (1932), 751–758.
Mulliken, R. S.: Quantum theory of the double bond, in: JACS 54 (1932), 4111–12.
Mulliken, R. S.: Electronic structures of polyatomic molecules and valence. IV. Electronic states, quantum theory of the double bond, in: Physical Review 43 (1933), 279–302.
Mulliken, R. S.: Intensities of electronic transitions in molecular spectra. I. Introduction, in: JCP 7 (1939), 14–20; II. Chargetransfer spectra, in: JCP 7 (1939), 20–34; III. Organic molecule with double bonds, in: JCP 7 (1939), 121–135; IV. Cyclic dienes and hyperconjugation, in: JCP7 (1939), 339–352; V. Benzene, in: JCP 7 (1939), 353–356; VI. Molecular refractivities of organic compounds, in: JCP 7 (1939), 356–363; VII. Conjugated polyenes and carotenoids, in: JCP 7 (1939), 364–373; VIII. Odd-numbered conjugated polyene chain molecules and organic dyes (with notes on optical anisotropy and Raman intensities), in: JCP 7 (1939), 570–572; IX. Calculations on the long wave-length halogen spectra, in: JCP 8 (1940), 234–243; X. Calculations on mixed-halogen, hydrogen halide, alkyl halide, and hydroxyl spectra, in: JCP 8 (1940), 382–395.
Mulliken, R. S.: Molecular scientists and molecular science: some reminiscences, in: JCP 43 (1965), S2–S11, S8.
Mulliken, R. S.: Life of a Scientist. Springer-Verlag, Berlin, Heidelberg, New York, London, Paris, Tokyo, 1989.
Munk, M., Hückel, E.: Systematische Messungen an Flügelprofilen, in: Mitteilungen der Göttinger Modellversuchsanstalt für Aerodynamik, Nr. 2, erschienen in Technische Berichte (1917), 148–163.
Munk, M., Hückel, E.: Weitere Untersuchungen von Flügelprofilen, in: Mitt. d. MVA, Nr. 6, erschienen in Technische Berichte (1917), 204–218; Weitere Göttinger Flügeluntersuchungen, in: Mitt. d. MVA, Nr. 17, erschienen in Technische Berichte (1918), 407–450.
Munk, M., Hückel, E.: Der Profilwiderstand von Tragflügeln. Eine Zusammenfassung der bisherigen Göttinger Flügelprofilmessungen, in: Mitt. d. MVA, Nr. 18, erschienen in Technische Berichte (1918), 451–461.
Neidlein, R.: Walter Hückel (1895–1973), in: CB 113 (1980), I–V.
Nernst, W. und Orthman, W.: Die Verdünnungswärme von Salzen bei sehr kleinen Kozentrationen, in: ZPC 135 (1928), 199–208.
Neus, J.: Aromatizität: Geschichte und mathematische Analyse eines fundamentalen chemischen Begriffs. Hyle Studies in History and Philosophy of Chemistry 2, Hyle Publications, Karlsruhe 2002. Internet: http://www.hyle.org/publications/books/neus
Nielsen, K. A., Kragh, H.: An Institute for Dollars: Physical Chemistry in Copenhagen between the world wars, in: Centaurus, 39 (1997), 311–331.
Nordheim, L.: Zur Quantentheorie des Wasserstoffmolekülions, in: ZP 19 (1923) 69–93.
Nye, M. J.: Physics And Chemistry: Commensurate Or Incommensurate Sciences?, in: The Invention of Physical Science, Intersections of Mathematics, Theology and Natural Philosophy

Since the Seventeenth Century – Essays in Honor of Erwin N. Hiebert. Kluwer Academic Publishers, Dordrecht, 1992.

Nye, M. J.: From Chemical Philosophy to Theoretical Chemistry: Dynamics of Matter and Dynamics of Disciplines, 1800–1950. University of California Press, Berkeley, CA, 1994.

Nye, M. J.: Remodeling a Classic: The Electron in Organic Chemistry, 1900–1940, in: Histories of the Electron, The Birth of Microphysics, edited by Jed Z. Buchwald and Andrew Warwick. The MIT Press, Cambridge, MA, 2001, pp. 339–361.

Nye, M. J.: Paper Tools and Molecular Architecture in the Chemistry of Linus Pauling, in: Tools and Modes of Representation in the Laboratory Sciences, Edited by Ursula Klein. Kluwer Academic Publishers, Dordrecht / Boston / London, 2001, Seite 117–132.

Onsager, L.: Zur Theorie der Elektrolyte. I, II, in: PZ 27 (1926), 388–392, 28 (1927), 277–298.

Pais, A.: Niels Bohr's Times: In Physics, Philosophy and Polity, Oxford University Press, Oxford, 1991.

Palmer, W. G.: A History of the Concept of Valency to 1930. Cambridge University Press, Cambridge, 1965.

Park, B. S.: Computations and Interpretations: The Growth of Quantum Chemistry, 1927–1967, Dissertation, Baltimore, Maryland, May 1999 (UMI Microfilm 9927141).

Park, B. S.: Chemical translators: Pauling, Wheland and their strategies for teaching the theory of resonance, in: British Journal for the History of Science 32 (1999), 21–46.

Park, B. S.: The contexts of simultaneous discovery: Slater, Pauling, and the origins of hybridisation, in: Studies in History and Philosophy of Science 31B (4) (2000), 451–474.

Pauli, W.: Über das Modell des Wasserstoffmolekülions, in: Annalen der Physik (4) 68 (1922), 177–240.

Pauling, L.: The calculation of matrix elements for lewis electronic structures of molecules, in: JCP 1 (1931), 280–283.

Pauling, L.: The nature of the chemical bond. Application of results obtained from the quantum mechanics and from a theory of paramagnetic susceptibility to the structure of molecules, in: JACS 53 (1931), 1367–1400.

Pauling, L.: The nature of the chemical bond. II. The one – electron bond and the three – electron bond, in: JACS 53 (1931), 3225–3237.

Pauling, L.: The nature of the chemical bond. III. The transition from one extreme bond type to another, in: JACS 54 (1932), 988–1003.

Pauling, L.: The nature of the chemical bond. IV. The energy of single bonds and the relative electronegativity of atoms, in: JACS 54 (1932), 3570–3582.

Pauling, L. and Sherman, J.: The nature of the chemical bond. VI. The calculation from thermochemical data of the energy of resonance of molecules among several electronic structures, in: JCP 1 (1933), 606–617; The nature of the chemical bond. VII. The calculation of resonance energy in conjugated systems, in: JCP1 (1933), 679–686.

Pauling, L. and Wheland, G.: The nature of the chemical bond. V. The quantum mechanical calculation of the resonance energy of benzene and naphthalene and the hydrocarbon free radicals, in: JCP 1 (1933), 362–374.

Pauling, L.: The diamagnetic anisotropy of aromatic molecules, in: JCP 4 (1938), 673–677.

Peyerimhoff, S.: Theoretische Chemie: Quantenchemie, in Forschung in der Bundesrepublik Deutschland, hrsg. Von Christoph Schneider, Verlag Chemie, Weinheim 1983, Seite 611–619.

Peyerimhoff, S.: The development of computational chemistry in Germany, in: Reviews in Computational Chemistry, Volume 18, Edited by Kenny B. Lipkowitz and Donald B. Boyd, Wiley – VCH, John Wiley and Sons, Inc. 2002.

Primas, V.: Müller-Herold, Elementare Quantenchemie, Teubner, Stuttgart 1984.

Psaros, N., Ruthenberg, K., Schummer, J.(Hrsg.): Philosophie der Chemie. Bestandsaufnahme und Ausblick. Königshausen & Neumann, Würzburg, 1996.

Pullman, B., Pullman, A.: Les théories électroniques de la chimie organique. Masson, Paris 1952.

Raman, C. V., Krishnan, K. S.: Magnetic double-refraction in liquids. Part I – benzene and its derivates, in: Proceedings of the Royal Society of London A 113 (1927), 511–519.

Ramberg, P. J.: Chemical Structure, Spatial Arrangement. The Early History of Stereochemistry, 1874–1914. Asgate, Aldershot 2003.

Rechenberg, H.: Die erste Göttinger Zeit und die Rostocker Professur (1919–1929), in: Hundert Jahre Friedrich Hund, Ein Rückblick auf das Wirken eines bedeutenden Physikers, hrsg. Von Manfred Schroeder, Nachrichten der Akademie der Wissenschaften in Göttingen, mathematisch-physikalische Klasse, Vandenhoek & Ruprecht in Göttingen, 1996.

Regitz, M., Heydt, H., Schank, K., Franke, W.: Bernd Eistert (1902–1978), in: CB 113 (1980), XXIX–LVIII.

Reid, C.: Hilbert. Springer-Verlag, Berlin, 1972.

Reinhard, C.: Forschung in der chemischen Industrie: Die Entwicklung synthetischer Farbstoffe bei BASF und Hoechst, 1863 bis 1914. Technische Universität Bergakademie Freiberg: TU Bergakademie, Freiberg, 1997.

Reinhardt, C.: Basic Research in Industry: Two Case Studies at I. G. Farbenindustrie AG in the 1920s and 1930s, in: Travis A. S., Schröter H. G., E. Homburg: Determinants in the Evolution of the European Chemical Industry, 1900–1939: New Technologies, Political Frameworks, Markets and Companies. Kluwer, Dordrecht, 1999.

Reinhardt, C., Travis, A. S.: Heinrich Caro and the Creation of Modern Chemical Industry. Kluwer Academic Publishers, Dordrecht, 2000.

Reinhardt, C. (Ed.): Chemical Sciences in the 20th Century: Bridging Boundaries. Wiley-VCH, Weinheim, 2001.

Richter, F.: Das Verhältnis der klassischen Stereochemie zu den neueren Arbeiten K. Weissenbergs, in: Naturwissenschaften, 14 (1926), 889–895.

Ritter, H., Zerweck, W.: Arthur von Weinberg, 1860–1943, in: Chemische Berichte 89 (1956), XIX–XLI.

Robertson, P.: The Early Years: The Niels Bohr Institut 1921–1930, Akademisk Forlag, Copenhagen, 1979.

Röseberg, U.: Niels Bohr: Leben und Werk eines Atomphysikers 1885–1962. Akademie-Verlag, Berlin 1987.

Rosenfeld, L.: Quantentheorie 1929: Erinnerungen an die erste Kopenhagenkonferenz, in: Niels Bohr 1885–1962, Der Kopenhagener Geist in der Physik, Friedr. Vieweg & Sohn, Braunschweig/Wiesbaden, 1985.

Rotta, J. C.: Die Aerodynamische Versuchsanstalt in Göttigen, ein Werk Ludwig Prandtls. Ihre Geschichte von den Anfängen bis 1925. Vandenhoeck & Ruprecht, Göttingen, 1990.

Rowe, D.: Einstein Meets Hilbert: At the Crossroads of Physics and Mathematics, in: Physics in Perspective 3 (2001), 379–424.

Russel, C. A.: The History of Valency. Humanities Press Inc., New York, 1971.

Ruedenberg, K.: The physical nature of the chemical bond, in: Review of Modern Physics 34 (1962), 326–376.

Rumer, G.: Zur Theorie der Spinvalenz, in: Nachrichten von der königlichen Gesellschaft der Wissenschaften zu Göttingen. Mathematisch-physikalische Klasse aus dem Jahre (1932), 337–341.

Rumer, G., Teller E., Weyl, H: Eine für die Valenztheorie geeignet Basis der binären Vektorinvarianten, in: Nachrichten von der königlichen Gesellschaft der Wissenschaften zu Göttingen. Mathematisch-physikalische Klasse aus dem Jahre (1932), 449–504.

Sachse, H.: Ueber die geometrischen Isomerien der Hexamethylenderivate, in: B 23 (1890), 1363–1370.

Sachse, H.: Über die Konfigurationen der Polymethylenringe, in: ZPC 10 (1892), 203–241.

Saltzman, M.: James Bryant Conant and the development of physical organic chemistry, in: Journal of Chemical Education 49 (1972), 411–412.

Saltzman, M.: The development of physical organic chemistry in the United States and the United Kingdom: 1919–1939, Parallels and Contrasts, in: Chemistry in Britain 63 (1986), 588–593.

Saltzmann, M.: C. K. Ingold's development of the concept of mesomerism, in: Bulletin for the History of Chemistry 19 (1996), 25–32.

Schleyer, P. v. R., Jiao, H.: What is aromaticity?, in: Pure & Applied Chemistry 68 (1996), 209–218.

Scholz, M., Köhler, H.:Quantenchemie – Quantenchemische Näherungsverfahren und ihre Anwendung in der organischen Chemie. VEB Deutscher Verlag der Wissenschaften, Berlin, 1981.

Schütt, Hans-Werner: Der Wandel des Begriffs "aromatisch" in der Chemie, in: Begriffswandel und Erkenntnisfortschritt in den Erfahrungswissenschaften. Kolloquium an der Technischen Universität Berlin, WS 84/85. Universitätsbibliothek der Technischen Universität Berlin, Berlin, 1985.

Schwarz, W. H. E., Karachalios, A., et al.: Hans G. A. Hellmann (1903–1938) I. Ein Pionier der Quantenchemie, in: Bunsen-Magazin 1 (1999), 10–21; II. Ein deutscher Pionier der Quantenchemie in Moskau, in: Bunsen-Magazin 1 (1999), 60–70.

Schwarz, W. H. Eugen: Messung von Orbitalen und Schrödinger-Katzen, in: Bunsen-Magazin 3 (2001), 59–68.

Schweber, S.: The Young John Clark Slater and the development of quantum chemistry, in: Historical Studies in the Physical and Biological Sciences 20 (1990), 339–406.

Servos, John W.: Physical Chemistry from Ostwald to Pauling, the Making of a Science in America. Princeton University Press, Princeton, NJ, 1990.

Siegmund-Schultze, Reinhard: Rockefeller and the Internationalization of Mathematics Between the Two World Wars. Birkhäuser Verlag, Basel, Boston, Berlin, 2001.

Simões, A.: Converging Trajectories, Diverging Traditions: Chemical Bond, Valence, Quantum Mechanics and Chemistry, 1927–1937, Ph. D. Thesis, University of Maryland, College Park, MD, 1993.

Simões, A.: Chemical physics and quantum chemistry in the twentieth-century, in: Cambridge History of Science (8 Vols.), Volume 5, Modern Physical and Mathematical Sciences, edited by Mary Jo Nye. Cambridge University Press, Cambridge 2003, pp. 394–412.

Simões, A., Gavroglu, K.: Different Legacies and Common Aims: Robert Mulliken, Linus Pauling and the Origins of Quantum Chemistry, in: Conceptual Perspectives in Quantum Chemistry, edited by Calais, J. L. and Kryachko, E. S.. Kluwer Academic Publishers, Dordrecht, 1997, 383–413.

Simões, A., Gavroglu, K.: Qunatum Chemistry *qua* Applied Mathematics. The Contributions of Charles Alfred Coulson (1910–1974), in: Historical Studies in the Physical Sciences 29 (1999), 363–406.

Simões, A., Gavroglu, K.: Quantum Chemistry in Great Britain: Developing a Mathematical Framework for Quantum Chemistry, in: Studies in the History and Philosophy of Modern Physics, 31B (2000), 511–548.

Simões, A., Gavroglu, K.: Issues in the History of Theoretical and Quantum Chemistry, 1927–1960, in: Chemical Sciences in the 20th Century. Bridging Boundaries, edited by Carsten Reinhardt. Wiley-VCH, Weinheim, 2001, pp. 51–74.

Slater, J. C.: The theory of complex spectra, in: Physical Review 34 (1929), 1293–1322.

Slater, J. C.: Cohesion in monovalent metals, in: Physical Review 35 (1930), 511–529

Slater, J. C.: Directed valence in polyatomic molecules, in: Physical Review 37 (1931), 481–489.

Slater, J. C.: Molecular energy levels and valence bonds, in: Physical Review 38 (1931), 1109–1114.

Slater, J. C.: Solid State and Molecular Theory: A Scientific Biography. John Wiley, New York, 1975.

Snelders, H. A. M.: The historical background of Debye and Hückel's theory of strong electrolytes, in: Proceeding of the Royal Netherlands Academy of Arts and Sciences B 89 (1986), 79–94.

Sobczyñska, D.: Fullerenes: The Philosophical Aspects of their Discovery, in: ARS MUTANDI. Issues in Philosophy and History of Chemistry, edited by Nikos Psarros and Kostas Gavroglu. Leipziger Universitätsverlag, Leipzig, 1999.

Staley, R.: Max Born and the German physics community: The education of a physicist. PhD thesis, Cambridge University, Cambridge (UK), 1992.

Stanley, A., Owens, G.: Energy and Conflict: The Life and Times of Edward Teller. G. P. Putnams´ Sons, New York, 1976.

Stocklöv, J.: Arthur Hantzsch: Wegbereiter der physikalischen organischen Chemie, Dissertation, Halle (Saale) 1996.

Stoermer, R. und Simon, M.: Ueber geometrisch isomere Derivate des Diphenyläthylens und deren Configuration, in: Liebig's Annalen der Chemie 342 (1905), 1–13.

Stoermer, R.: Über die Umlagerung stabiler stereoisomerer Äthylenkörper in labile durch ultraviolettes Licht (I), in: B 42 (1909), 4865–4871; Über die Umlagerung der stabiler stereoisomerer in labile durch ultraviolettes Licht (II), in: B 44 (1911), 637–668.

Stoemer, R.: Hrn. Liebermann zur Erwiderung. – Zur Kenntnis der halogenierten Zimtsäuren und deren Verhalten im ultravioletten Licht, in: B 46 (1913), 1249–1266.

Stoermer, R., Heymann, P.: Die Bestimmung der Konfiguration der stereoisomeren Zimtsäuren, in: B 45 (1912), 3099–3104.

Stokes, R. G.: Von der I. G. Farbenindustrie AG bis zur Neugründung der BASF (1925–1952), in: Die BASF. Eine Unternehmensgeschichte, Verlag C. H. Beck, München, 2002.

Stranges, A. N.: Electrons and Valence: Development of the Theory, 1900–1925. Texas A&M University Press, Texas, 1982.

Streitwieser, A.: Molecular Orbital Theory for Organic Chemists. Wiley, New York 1961.

Streitwieser, A.: A Lifetime of Synergy with Theory and Experiment, Jeffrey I. Seeman, Series Editor, American Chemical Society, Washington, DC, 1997.

Streitwieser, A.: History of Computational Chemistry: A Personal View, in: Encyclopedia of Computational Chemistry, vol 2, Seite 1237–1243.

Suchy, K.: Erich Hückel zum Gedenken, in: Physikalische Blätter 36 (1980), 248–249.

Suchy, K.: Erich Hückel, in: Physics Today 33 (1980), 72–75.

Sutton, L. E.: The significance of the differences between the dipole moments of saturated and unsaturated substances, in: Proceedings of the Royal Society of London (A) 133 (1931), 668–695.

Sylvester, J. J.: On the application of the new atomic theory to the graphical representation of the invariants and covariants of binary quantics, in: American Journal of Mathematics 1 (1878), 64–90.

Trischler, H.: Luft- und Raumfahrtforschung in Deutschland 1900–1970: Politische Geschichte eines Wissenschaft. Frankfurt am Main, Campus (1992).

Üstün, A.: Zweites Vaterland – deutsche Chemiker im türkischen Exil, in: Nachrichten aus der Chemie 51 (2003), 152–155.

van't Hoff, J. H.: La chimie dans l`espace. Bazendijk, Rotterdam, 1875.

Voigt, J. H.: Die Universität Stuttgart. Phasen ihrer Geschichte. Wittwer, Stuttgart, 1981.

Vorländer, D.: Ueber krystallinisch-flüssige Substanzen, in: B 39 (1906) 803–810.

Vorländer, D.: Einfluß der molekularen Gestalt auf den krystallinisch-flüssigen Zustand, in: B 40 (1907) 1970–1972.

Vorländer, D.: Krystallinisch-flüssige Substanzen. F. Enke, Stuttgart, 1908.

Vorländer, D., Kasten, W.: Ueber durchsichtig klare krystallinische Flüssigkeiten, in: B 41 (1908) 2033–2052.

Vorländer, D., Siebert, E., Spreckels, E.: Die Lehre von den innermolekularen Gegensätzen und die Theorie des Benzols, in: B. 52 (1919), 263–283.

Vorländer, D.: Die Lehre von den innermolekularen Gegensätzen und die Lenkung der Substituenten im Benzol II., in: B. 58 (1925), 1893–1914.

Vorländer, D.: Säure, Salz, Ester und Addukt, in: B 58 (1925), 118–143.

Vorländer, H.: Die NSV: Darstellung und Dokumentation einer nationalsozialistischen Organisation. Harald Boldt Verlag, Boppard am Rhein, 1988.

Wachholtz, F.: Die Umwandlung von Malein- in Fumarsäure durch Bromatome, in: ZPC 135 (1928), 147–181.

Walter, W., Eistert, B:. Fritz Arndt (1885–1969), in: CB 108 (1975), I-XLIV.

Weyer, J.: Joseph Achille Le Bel 1847–1930, in: Chemie in unserer Zeit 8 (1974), 143–158.

Weyer, J.: Hundert Jahre Stereochemie – Ein Rückblick auf die wichtigsten Entwicklungsphasen, in: Angewandte Chemie 86 (1974), 604–611.

Wheland, G. W., Pauling, L.: A quantum mechanical discussion of orientation of substituents in aromatic molecules, in: JACS 57 (1935), 2086–2095.

Weinberg, A. von: Über organische Doppelbindungen, in: Zeitschrift für angew.andte Chemie 44 (1931), 815–818.

Weinberg, A. von: Zum Benzol – Problem, in: B. 52 (1919), 928–940; Zum Benzol – Problem (II), in: B. 52 (1919), 1501–1508; Zum Benzol – Problem (III), in: B. 53 (1920), 1353–1358; Zum Benzol – Problem (IV): Die Naphthalinformel, in: B. 54 (1921), 2168–2171; Zum Benzol-Problem V: Der Benzolring in Substitutionsprodukte, in: B. 54 (1921), 2171–2175.

Weininger, S. J.: Affinity, additivity and the reification of the bond, in: Tools and Modes of Representation in the Laboratory Sciences, Edited by Ursula Klein. Kluwer Academic Publishers, Dordrecht / Boston / London, 2001, 237–251.

Weyl, H.: Zur quantentheoretischen Berechnung molekularer Bindungsenergien, in: Nachrichten von der königlichen Gesellschaft der Wissenschaften zu Göttingen. Mathematisch-physikalische Klasse aus dem Jahre, 1930, 285–294.

Weyl, H.: Zur quantentheoretischen Berechnung molekularer Bindungsenergien. II, in: Nachrichten von der königlichen Gesellschaft der Wissenschaften zu Göttingen. Mathematisch-physikalische Klasse aus dem Jahre, 1931, 33–39.

Weyngard, C.: Daniel Vorländer, in: B 76 (1943), 41–58.

Wigner, E.: An Appreciation on the 60th Birthday of Edward Teller, in: H. Mark, Fernbach (Hrsg.): Properties of Matter Under Unusual Conditions. In Honor of Edward Teller`s 60th Birthday. New York, 1969.

Willstätter, R., Waser, E.: Über Cyclo-octatetraen, in: B. 44 (1911), 3423–3445.

Willstätter, R., Heidelberger, M.: Zur Kenntnis des Cyclo-octatetraens, in: B. 46 (1913), 517–527.

Wislicenus, J.: Über die räumliche Anordnung der Atome in organischen Molekülen. Leipzig 1887.

Wolfenden, J. H.: The anomaly of strong electrolytes, in: Ambix 19 (1972), 175–196.

Woodward, R. B; Hoffmann, R.: Die Erhaltung der Orbitalsymmetrie. VCH Verlagsgesellschaft, Weinheim 1970.

Yates, K.: Hückel Molecular Orbital Theory. Academic Press, New York, 1978.

Zsigmondy, R., Hückel, E.: Über Reduktionsgeschwindigkeit und das Wachstum kleiner Goldteilchen bei der Herstellung kolloidaler Goldlösungen, in: ZPC 116 (1925), 291–303.

Index

Kolibri 360 GmbH
Ferdinand-Jühlke-Straße 7
99095 Erfurt, Deutschland
produktsicherheit@kolibri360.de